Nuclear Milestones

At the Atomic Pioneer Award ceremony: (left to right) Dr. Glenn T. Seaborg, President Richard M. Nixon, Dr. James B. Conant, Gen. Leslie Groves, and Dr. Vannevar Bush.

Nuclear Milestones

A COLLECTION OF SPEECHES BY

Glenn T. Seaborg
University of California
Berkeley, California

W. H. FREEMAN AND COMPANY
San Francisco

Portions of this book have been released
previously by the U.S. Atomic Energy Commission.

Printed in the United States of America

International Standard Book Number: 0-7167-0342-4

Library of Congress Catalog Card Number: 72-3915

To

Theos J. Thompson (1918-1970)
Commissioner,
U.S. Atomic Energy Commission (1969-1970)

and

Burris B. Cunningham (1912-1971)
Nuclear Chemist Extraordinaire

Contents

Preface

In our travels into the Nuclear Age over the past quarter of a century, we have passed many milestones of discovery and development. During these years we have seen new fields of science born and grow, new national laboratories brought to life and maturity, and new technologies developed and applied to advance mankind.

Along the way we have paused to recognize these accomplishments, to take note of their origins, and to evaluate where they may be leading us. Celebrations of anniversaries and other events often give us a chance to reminisce and reflect on these accomplishments. And in this book, through a selection of words spoken at those celebrations during my tenure as Chairman of the U.S. Atomic Energy Commission, I try to present some of those reminiscences and reflections as well as to reveal some personal and scientific history.

The first part, Builders and Discoverers, focuses on many of the people who were instrumental in pioneering the Nuclear Age. It tells the story of men whose work rarely made headlines but whose accomplishments led to the challenging era in which we live.

The second part, Historic Landmarks, has its major focus on the institutions in which the great discoveries and developments of nuclear science and technology were made and advanced. Many of these institutions continue in their role as nuclear pioneers, probing new frontiers of the atom and helping the nation to use its nuclear knowledge and power for peace and human progress and, in an expanded role, exploiting science and technology on a broader basis for these purposes.

May 1972

Glenn T. Seaborg

PART ONE

Builders and Discoverers

307 GILMAN HALL

Plutonium, element 94, was discovered in this room on the night of Feb. 23, 1941.

. . . Some Reminiscences

At the dedication of Room 307 Gilman Hall as a National Historic Landmark, University of California, Berkeley, California, Feb. 21, 1966

■ I am happy to participate with Arthur Wahl and Edwin McMillan in this twenty-fifth anniversary of the discovery of plutonium and in the dedication of Room 307 Gilman Hall as a National Historic Landmark.

I imagine it is typical of our times—because of the speed of change, the sheer number of significant events that pile up in the quickly passing years—that each of us today lives through a little more history in our lifetime.

At least this seems to be the case. It is an exciting time to be alive, to be working, to be trying to make some contribution to the scheme of things, and occasionally to have some small success in the effort.

It is also a time when time itself is something of a luxury—particularly time to reminisce. But, since today is a special occasion, I hope you will grant me a little of that luxury.

Having a room in which you and your colleagues worked rather routinely, and certainly unceremoniously, designated as a National Historic Landmark is an unusual experience, to say the least. Those of you who remember Room 307 Gilman Hall as it was in those early days (and as it remained for many years) will agree that a less significant or historical looking room hardly existed on the campus of the University of California.

Fortunately the room is still here. It has been enlarged somewhat, and it contains more complicated equipment. The simple small sink, down which some of our precious plutonium was inevitably lost in

Fortunately the room is still here. . .

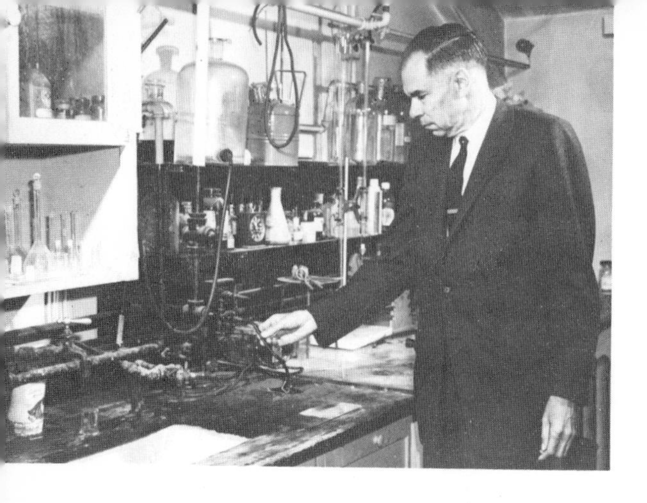

Glenn T. Seaborg visiting the Room 307 laboratory where he and his colleagues discovered plutonium.

the course of our experiments 25 years ago, has been replaced by another sink. But the little cubbyhole with its low slanting ceiling directly under Gilman Hall's roof, where we kept our electroscope and various samples, is still an appendage to the room. And it still opens through glass doors to the little outdoor patio where, because of the shortage of laboratory space and fume hoods, we were forced to carry on some of our experiments that gave off noxious fumes.

I recall that our counting equipment was two doors down the hall, in Room 303. The alpha radiation from the first plutonium was measured in that room, and therefore Room 303 shares a place in history with Room 307. Joseph Kennedy and I had our desks in Room 303, and later, in 1941, one whole wall was taken up with a Chart of Isotopes to which additions and changes were frequently made.

Had Art Wahl, Joe Kennedy, Ed McMillan, or I had the slightest idea that today's event would transpire, we might have looked for more auspicious quarters. I do not think we would have gotten

Room 307 still opens to the little outdoor patio. . .

Glenn T. Seaborg, Arthur C. Wahl, and Edwin M. McMillan on the patio off Room 307.

them. Space was at a premium, and we were lucky to have even these rooms to work in. Fortunately we were more interested in getting results in our work than in our surroundings or any significance they might have in the future.

In recalling the story of plutonium, I should go back further, perhaps to 1936, when as a graduate student I spoke in the College of Chemistry weekly seminar, as was required of each of us once a year. Since the fall of 1934, when I began my graduate work at Berkeley, I had been reading first the exciting papers by Fermi, Segré, and coworkers from Rome and then the equally fascinating papers by Hahn, Meitner, and Strassmann from Berlin. They were studying the interesting radioactivities which were produced when uranium was bombarded with neutrons and which they attributed to isotopes of transuranium elements. I remember how I devoured those early papers and how I considered myself something of a minor expert on the "transuranium elements." In fact, they were the subject of my talk at that seminar in 1936, an hour-long talk in which I described those "new" elements and their chemical properties in great detail. I need not remind you, I am sure, that in January 1939 word reached us that Hahn and Strassmann in Germany had identified the transuranium isotopes as barium, lanthanum, and other fission products of uranium and thus established that they were not new elements at all.

During the two years following my seminar talk in 1936 and before the discovery of fission, my interest in the neutron-induced radioactivities in uranium continued unabated and, in fact, increased. I read and reread every article published on the subject. I was puzzled by the situation, both intrigued by the concept of the transuranium interpretation of the experimental results and disturbed by the apparent inconsistencies in this interpretation. I

Joseph W. Kennedy.

On this exciting night in January 1939, we heard the news from Germany of Hahn and Strassmann's beautiful chemical experiments. . .

---

remember discussing the problem with Joe Kennedy by the hour, often in the postmidnight hours of the morning at the old Varsity Coffee Shop on the corner of Telegraph and Bancroft Avenues where we often went for a cup of coffee and a bite to eat after an evening spent in the laboratory.

I first learned of the correct interpretation of these experiments, that neutrons split uranium into two large pieces in the fission reaction, at the weekly Monday night seminar in nuclear physics conducted by Professor E. O. Lawrence in Le Conte Hall. On this exciting night in January 1939, we heard the news from Germany of Hahn and Strassmann's beautiful chemical experiments. I recall that at first the fission interpretation was greeted with some skepticism by a number of those present, but, as a chemist with a particular appreciation for Hahn and Strassmann's experiments, I felt that this interpretation just had to be accepted. I remember walking the streets of Berkeley for hours after this seminar in a combined state of exhilaration in appreciation of the beauty of the work and of disgust at my inability to arrive at this interpretation despite my years of contemplation on the subject.

With those radioactivities identified as fission products, there were no longer any transuranium elements left. However, in later investigations by Ed McMillan at Berkeley and others elsewhere, one of the radioactivities behaved differently from the others. The beta radioactivity with a half-life of about 2.3 days did not undergo recoil. It did not separate from thin layers of uranium when uranium was bombarded with slow neutrons. Along toward the spring of 1940, Ed began to come to the conclusion that the 2.3-day activity might actually be due to the daughter of the 23-minute uranium-239 and thus might indeed be an isotope of element 93 with the mass number 239 (93-239). Phil Abelson joined him in this work in the spring of 1940, and together they were able to chemically separate and identify and thus discover element 93.

Immediately thereafter, during the summer and fall of 1940, Ed McMillan started looking for the daughter product of the 2.3-day activity, which obviously would be the isotope of element 94 with mass number 239 (94-239). Not finding anything he could positively identify as such, he began to bombard uranium with deuterons in the 60-inch cyclotron in the hope that he might find a shorter lived isotope—one of a higher intensity of radioactivity that would be easier to identify as an isotope of element 94. Before he could finish this project, he was called away to work on radar at M.I.T.

During this time my interest in the transuranium elements continued. Since Ed McMillan and I lived only a few rooms apart in the Faculty Club, we saw each other quite often, and, as I recall, much of our conversation, whether in the laboratory, at meals, in the hallway, or even going in and out of the shower, had something to do with element 93 and the search for element 94. I must say, therefore, that his sudden departure for M.I.T. came as something of a surprise to me—especially since I did not even know when he had left.

In the meantime, I asked Arthur Wahl, one of my two graduate students, to begin studying the tracer chemical properties of element 93 with the idea that this might be a good subject for his thesis. My other coworker was Joe Kennedy, a fellow instructor at the University and, as I have indicated, also very interested in the general transuranium problem.

When I learned that Ed McMillan had gone, I wrote to him asking whether it might not be a good idea if we carried on the work he had started, especially the deuteron bombardment of uranium. He readily assented.

Our first deuteron bombardment of uranium was conducted on Dec. 14, 1940. What we bombarded was a form of uranium oxide, U_3O_8, which was literally plastered onto a copper backing plate. From this bombarded material Art Wahl isolated a chemical fraction of element 93. The radioactivity of this fraction was measured and studied. We observed that it had different characteristics than the radiation from a sample of pure 93-239. The beta particles, which in this case were due to a mixture of 93-239 and the new isotope of element 93 with mass number 238 (93-238), had a somewhat higher energy than the radiation from pure 93-239 and there was more gamma radiation. But the composite half-life was about the same, namely, 2 days. However, the sample also differed in another very important way from a sample of pure 93-239. Into this sample there grew an alpha-particle-emitting radioactivity. A proportional counter was used to count the alpha particles to the exclusion of the beta particles. This work led us to the conclusion that we had a daughter of the new isotope 93-238—a daughter with a half-life of about

Our experiments gave us proof that what we had made was chemically different from all other known elements. . .

---

50 years and with the atomic number 94. This is much shorter lived than the now known half-life of 94-239, which is 24,000 years. The shorter half-life means a higher intensity of alpha-particle emission, which explains why it was so much easier to identify what proved to be the isotope of element 94 with the mass number 238 (94-238). (Later it was proved that the true half-life of what we had, i.e., 94-238, is about 90 years.)

On Jan. 28, 1941, we sent a short note to Washington describing our initial studies on element 94; these data also served for later publication in *The Physical Review* under the names of McMillan, Wahl, Kennedy, and Seaborg. We did not consider, however, that we had sufficient proof at that time to say we had discovered a new element and felt that we had to have chemical proof to be positive. So, during the rest of January and into February, we attempted to identify this alpha activity chemically.

Our attempts proved unsuccessful for some time. We did not find it possible to oxidize the isotope responsible for this alpha radioactivity. Then I recall that we asked Professor Wendell Latimer, whose office was on the first floor of Gilman Hall, to suggest the strongest oxidizing agent he knew for use in aqueous solution. At his suggestion we used peroxydisulphate with argentic ion as catalyst.

On the stormy night of Feb. 23, 1941, in an experiment that ran well into the next morning, Art Wahl performed the oxidation which gave us proof that what we had made was chemically different from all other known elements. That experiment, and hence the first chemical identification of element 94, took place in Room 307 of Gilman Hall, the room that is being dedicated as a National Historic Landmark today, 25 years later.

The communication to Washington describing this oxidation experiment, which was critical to the discovery of element 94, was sent on Mar. 7, 1941, and this served for later publication in *The Physical Review* under the authorship of Kennedy, Wahl, and Seaborg.

Almost concurrent with this work was the search for, and the demonstration of the fission of, the isotope of major importance—

The first tiny bit of man-made plutonium-239, with which its fission properties were established on Mar. 28, 1941, photographed in Room 307 on Feb. 21, 1966. This sample had recently been found where Glenn T. Seaborg placed it nearly twenty-five years before in a cigar box with other mementoes for safe-keeping. (The plutonium is too small an amount to be visible in the rare earth carrier material.)

94-239, the radioactive decay daughter of 93-239. Emilio Segrè played a major role in this work together with Kennedy, Wahl, and me. The importance of element 94 stems from its fission properties and its capability of production in large quantities. This, of course, is a story of more than Room 307 Gilman Hall. It involves, in addition, the 60-inch cyclotron, the Old Chemistry Building, the Crocker Laboratory, and the 37-inch cyclotron, all of which have by now been removed from the Berkeley campus. The 0.5-microgram sample on which the fission of 94-239 was first demonstrated was produced by transmutation of uranium with neutrons from the 60-inch cyclotron; it was chemically isolated in rooms in Old Chemistry Building and Crocker Laboratory and in Room 307 Gilman; and the fission counting was done using the neutrons from the 37-inch cyclotron.

How element 94 eventually got the name plutonium is an interesting story and one worth telling on this occasion. The work was carried on under self-imposed secrecy in view of its potential implications for national security. Following the discovery in February 1941 and well into 1942, we continued, as I have in my talk thus far, to use only the name "element 94" among ourselves

This, of course, is a story of more than Gilman Hall .

Below: The old Radiation Laboratory in the foreground. The Crocker Laboratory, where the 60-inch cyclotron was housed, is across the walk on the right. All buildings have been demolished. Right, top: Philip H. Abelson in 1940, when he and Edwin M. McMillan discovered element 93, neptunium. Right, middle: The deuteron beam of the 60-inch cyclotron. Right, bottom: Glenn T. Seaborg and Emilio Segré, on March 28, 1966, presenting to the Smithsonian Institution the historic 0.5-microgram sample of plutonium which had been placed in the cigar box 25 years before.

and the other few people who knew of the element's existence. But we needed a code name to be used when we might be overheard. Someone suggested "silver" as a code name for element 93, and we decided to use "copper" for element 94. This worked just fine until, for some reason I cannot recall now, it became necessary to use real copper in our work. Since we continued to call element 94 "copper," on occasion we had to refer to the real thing as "honest-to-God copper."

The first time a true name for element 94 seemed necessary was in writing the report to the Uranium Committee in Washington in March of 1942. I remember very clearly the debates within our small group as to what that name should be. It eventually became obvious to us that we should follow the lead of Ed McMillan, who had named element 93 neptunium because Neptune is the next planet after Uranus, which had served as the basis for the naming of uranium 150 years earlier. Thus we should name element 94 for Pluto, the next planet beyond Neptune. But, and this is a little known story, it seemed to us that one way of using the base Pluto was to name the element "plutium." We debated the question of whether the name should be "plutium" or "plutonium," the sound of which we liked much better. We finally decided to take the name that sounded better. I think we made a wise choice, and I believe it is also etymologically correct.

There was also the matter of the need for a symbol. Here, too, a great deal of debate was engendered because, although the symbol might have been "Pl," we liked the sound of "Pu"—for the reason you might suspect. We decided on "Pu," and, I might add, we expected a much greater reaction after it was declassified than we ever received.

Remembering the early days of the discovery and reading some of the early reports and correspondence brings to mind other events and thoughts that make one realize how times have changed. I recall reading the Uranium Committee report Phil Abelson wrote concerning the importance of our experiments. In it he said:

> Obviously, the results of these experiments will have a large bearing in the determination of the value of uranium power. It is probable that the cost of isotope separation will be great. The decision to spend perhaps a *million dollars* on a separation plant may well hinge on the results of these experiments.

A million dollars? The amount seemed astronomical to us at the time. We had no idea that our work would play a major role in a program that would eventually cost more than *two billion dollars* within a few years.

Only a short time after that Uranium Committee report, I recall that, at E. O. Lawrence's suggestion, I wrote to Lyman Briggs requesting a contract for some further work that might be done on the measurement of fission cross sections in the uranium and transuranium regions. Among the items listed on my proposal was a request for an assistant, a chemist Ph.D., at the tempting annual salary of $3,000! Another item on that request was the use of the 60-inch cyclotron at a cost of $25 per hour.

But far more dramatic than these personal recollections, and certainly far more important, was the development of plutonium starting with that first identification in 307 Gilman Hall. No other element has seen a similar growth. Our first experiments were done with tracer amounts as small as a picogram (a million millionth of a gram). Within five years our country was producing plutonium in kilogram amounts. The intervening twenty years have seen the production of somewhere between megagram and gigagram amounts—an escalation of a billion billion fold!

What is the significance of the growth of plutonium? What bearing will this element have on our future? As is true with all the power of science and technology at our command today, what will come of plutonium depends on how we (all mankind) choose to use it. I will not dwell on its destructive potential. This is well known by most of us here today and most people around the world—so much so that to many the symbol of the nuclear age is unfortunately one of horror. Perhaps fear of massive destruction will be the deciding factor that will bring men to choose reason rather than conflict in settling their differences. In this case, perhaps it will also be the power of plutonium, used constructively and beneficially, that will help men achieve some of the things essential to world stability, a more widely shared abundance, and a lasting peace.

Let me leave you with this concluding thought: The advent of plutonium, with its potential for war or peace, its possible use for the devastation of this planet or the lifting of all men to new standards of living, sharply brings into focus the major dilemma of our day. Can man, who now holds his destiny in his own hands, act with enough wisdom, patience, and understanding to choose the right path? I believe he can. I believe he will. I know that the years ahead will add strength to this conviction, and I hope that they will give cause to those who pass by Room 307 to stop now and then, recall what took place there, and perhaps recognize the event as one of the small but rewarding moments in a history leading us all to a better and brighter tomorrow. ■

25th Anniversary

THE FIRST WEIGHING OF PLUTONIUM

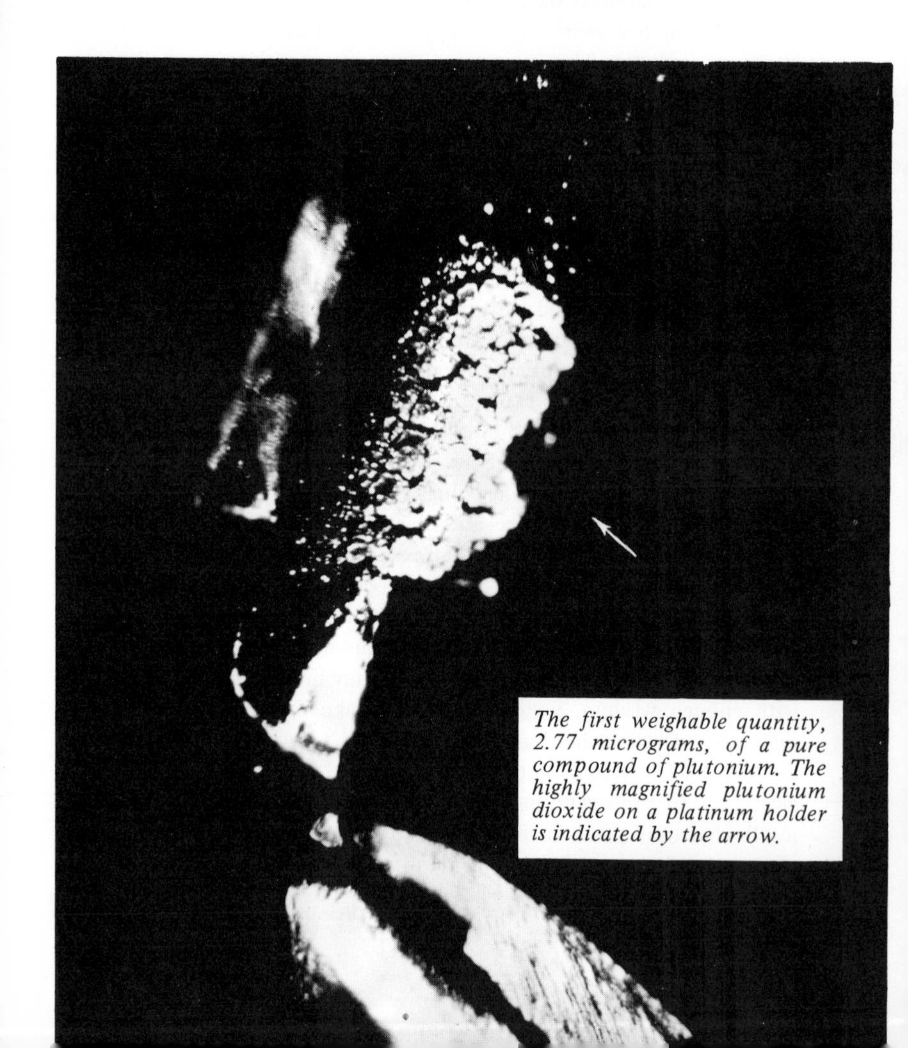

The first weighable quantity, 2.77 micrograms, of a pure compound of plutonium. The highly magnified plutonium dioxide on a platinum holder is indicated by the arrow.

Louis B. Werner, Glenn T. Seaborg, Burris B. Cunningham, and Michael Cefola (left to right) in Room 405 Jones Laboratory at the University of Chicago on Sept. 10, 1967, the twenty-fifth anniversary of the first weighing of plutonium.

On Sept. 10, 1967, a group of scientists held a reunion to celebrate the twenty-fifth anniversary of an important scientific event and to be present when the room where the event took place was designated as a National Historic Landmark.

The event, which took place on Sept. 10, 1942, was the first weighing of plutonium. The room where it took place was Room 405 Jones Laboratory at the University of Chicago, one tiny room of the wartime Metallurgical Laboratory of the Manhattan Project.

Two previous experiments had proved that plutonium-239 was even more fissionable than uranium-235, and thus it appeared possible that either of these isotopes might serve as the basic ingredient for a nuclear weapon. These crucial experiments were conducted on Mar. 28 and May 18, 1941, at Berkeley. The mission of the Met Lab was to develop (1) a method for the production of plutonium in quantity and (2) a method for its chemical separation on a large scale.

The key to solving the first problem was the demonstration by

---

We saw this first man-made element on Aug. 20, 1942 . . .

The first weighing of a 2.77-microgram sample was a key step in developing a method for its chemical separation on a large scale. . .

Enrico Fermi and his colleagues of the first sustained nuclear chain reaction on Dec. 2, 1942.

Important to the solution of the second problem was the determination of the chemical properties of plutonium, an element so new that little was known about its characteristics. The work was done with the extremely limited quantities available, and this first weighing of a 2.77-microgram sample was a key step in a crucial aspect of the project.

Solution of these two critical problems led to the construction of the large plutonium production reactors and chemical separation plants in Hanford, Washington, and to the success of a program that culminated with the detonation of a plutonium weapon over Nagasaki, which brought World War II to an end.

Plutonium still serves as a major ingredient of nuclear weapons, but perhaps more significant today is its peaceful potential as the fuel for the "breeder" type nuclear power reactor. Nuclear power stations using the breeder reactors now under development may someday be our main source of electricity, capable of supplying most of the world's growing power needs for centuries to come.

The following text, transcribed from the remarks of those scientists who gathered at the University of Chicago on Sept. 10, 1967, to celebrate the twenty-fifth anniversary of the first weighing of plutonium, tells an important part of the story of this fascinating new element that is destined to play an increasingly significant role in the future of man.

GLENN T. SEABORG The first weighing of plutonium was a significant event in the history of science and technology. A number of scientists assembled here in the Metallurgical Laboratory at the University of Chicago in the spring of 1942. Among those were a group of chemists working with me on a process to extract plutonium from uranium and its fission products. The uranium would be irradiated in huge production reactors at some site not yet chosen.

It occurred to me that central to the achievement of such a

separation process would be chemical work on concentrations that would exist in the chemical extraction plant. This seemed a very far-out idea, and I can remember a number of people telling me that they thought it was essentially impossible because we had no large source for plutonium. But I thought we could irradiate large amounts of uranium with the neutrons from cyclotrons since the indications were that we probably could produce sufficient plutonium, *if* we could learn to work on the microgram or smaller-than-microgram scale. That way we could get concentrations as large as those that would exist in the chemical extraction plant.

I knew rather vaguely about two schools of ultramicrochemistry—the School of Benedetti Pichler at Queens College in New York and the School of Paul Kirk in the Department of Biochemistry at the University of California at Berkeley.

I went to New York in May 1942, looked up Benedetti Pichler, and told him that I needed a good ultramicrochemist. He introduced me to Michael Cefola, and I offered him a job, which he accepted immediately. That he was on the job about three weeks later illustrates the pace at which things moved in those days.

Then, early in June, I took a trip to Berkeley, where I looked up my friend Paul Kirk and put the same problem to him. By the way, I could not tell any of these people why we wanted to work with microgram amounts or what the material was, but this did not seem to deter their willingness to accept. Paul Kirk introduced me to Burris Cunningham. When I asked him if he would come to Chicago, he accepted and was in town by the end of the month. He told me as soon as he arrived that he had a fine student, Louis Werner, he would like to invite, and I was, of course, delighted. Werner came along in a few weeks.

These, then, are the people who began the task of isolating plutonium from large amounts of uranium. We brought from Berkeley a little cyclotron-produced sample prepared by Art Wahl. It contained a microgram or so of plutonium mixed with several milligrams of rare earths. Using that sample, the ultramicrochemists, Cunningham, Cefola, and Werner, isolated the first visible amount—about a microgram—of pure plutonium. I guess it was a fluoride or perhaps a hydroxide. It was not weighed, but it could be seen! We were all very excited when we saw this first man-made element on Aug. 20, 1942.

In the meantime, hundreds of pounds of uranium were being bombarded with neutrons produced by the cyclotron at Washington University, under the leadership of Alex Langsdorf, and at the 60-inch cyclotron at Berkeley, under the leadership of Joe Hamilton.

This highly radioactive material was then shipped to Chicago. Art Jaffey, Truman Kohman, and Isadore Perlman led a team of chemists who put this material through the ether extraction process and the oxidation and reduction cycles to bring it down to a few milligrams of rare earths containing perhaps 100 micrograms of plutonium. This was turned over to Cunningham, Werner, and Cefola. These men prepared the first sample in pure form by going through the plutonium iodate and the hydroxide, etc., on to the oxide.

This 2.77-microgram sample was weighed on Sept. 10, 1942. The first aim was to weigh it with a so-called Emich balance, which was somewhat complicated and had electromagnetic compensation features. As it turned out, owing to the heavy load in the shops, this weighing balance would have taken perhaps six months to build.

Cunningham then had the idea of using a simple device consisting of a quartz fiber about 12 centimeters long and 1/10 of a millimeter

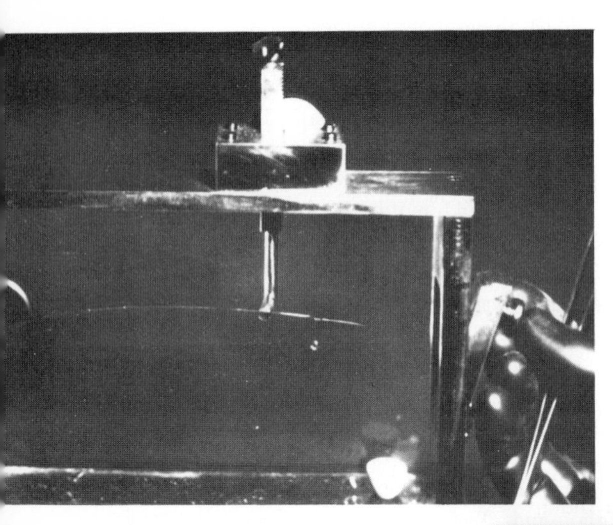

The Salvioni balance.

A simple device consisting of a quartz fiber. . .

Burris B. Cunningham operates the balance.

in diameter suspended at one end with a weighing pan hung on the other end. Then the depression of that end of the fiber with the pan containing the sample would relate to the weight of the sample. Cunningham measured the depression of the quartz fiber with a microscope. He built this balance himself, although he found out later that an Italian named Salvioni had invented it earlier, and so it became known as the Salvioni balance.

I would now like to call on Burris B. Cunningham for his recollections about the first weighing of plutonium.

BURRIS B. CUNNINGHAM As I look at this assemblage, I cannot resist remarking that rarely have so many made so much about so little.

The weighing experiments were only a part of a broad program of research on the properties of plutonium carried out under Glenn Seaborg's direction in the summer of 1942.

The experiments that immediately preceded the weighings and the weighings themselves represented two important scientific "firsts." They afforded the first human glimpse of a man-made element, and they were, to the best of my knowledge, the first ultramicro, gravimetric chemical experiments carried out in the United States.

Now, after all these years, it is difficult to recall the psychological impact of these events. Today alchemy is a thriving, commonplace business. But at that time we, who had been brought up in an older tradition, saw it as a miracle and just a little bit difficult to believe in.

More than a year after the first isolation of plutonium, I recall one of the members of our group arguing vehemently that plutonium was not really plutonium at all; it was just an oddly behaving isotope of uranium. The ultramicro work met with similar skepticism. When I first showed Dr. Seaborg the data on the reproducibility of the balance that we intended to use for weighing the plutonium, he thought that I had slipped a couple of decimal places and that these deviations must surely be in micrograms rather than in hundredths of a microgram. And I recall a long conversation with Truman Kohman in which I vainly tried to convince him that it was possible to measure a microliter of solution to 1% accuracy. I am not sure that he believes it even to this day.

Mike and Louie and I believed in plutonium but wondered constantly if the stuff we were precipitating from our little cones was genuinely pure material. There was always the possibility that it might be grossly contaminated with other material.

It was very doubtful whether you could even separate or purify and handle materials as small as micrograms and submicrograms. . .

And, of course, everybody worried about the calibration of the balance. How could you calibrate a balance to a hundredth of a microgram when you did not have microgram weights to do it with? These doubts dissipated gradually, and we came to accept the obvious. Plutonium *was* a little complicated in its chemical behavior perhaps, but it was much easier to purify than many elements that were discovered a half century earlier. And, after we calibrated the balance in three independent ways and came out with the same answer, we realized that Bureau of Standards certified microgram weights were not essential.

Looking back on these early experiments, one sees that they were not really glamorous or high-flown at all. They were straightforward and pretty simpleminded really. And, in a way, this seems a pity, because one feels that an event of such historical importance should have involved at least one esoteric principle of chemistry or physics. The balance ought really to have been something much more complicated than a quartz glass fiber enclosed in a homemade case of wood and glass. But that's the way it was. I suppose that most of us looking back on those early days and recalling the challenge and excitement would not change them if we could. I became hooked. I have since had the great pleasure of being the first to do experiments with other new elements. And, in a way, I am still doing business at the same old store. Only the atomic numbers have changed.

GLENN T. SEABORG It seems rather commonplace now that it should have been easy to handle materials on this scale. But at that time, it was very doubtful in our minds whether you could even separate or purify and handle, as a chemical entity, materials as small as micrograms and submicrograms.

Now I would like to call on Michael Cefola.

MICHAEL CEFOLA My initial knowledge of the existence of the project came when Dr. Seaborg contacted me in May 1942 to discuss my joining his group at the Metallurgical Laboratory.

I learned very little from the meeting about the nature of his

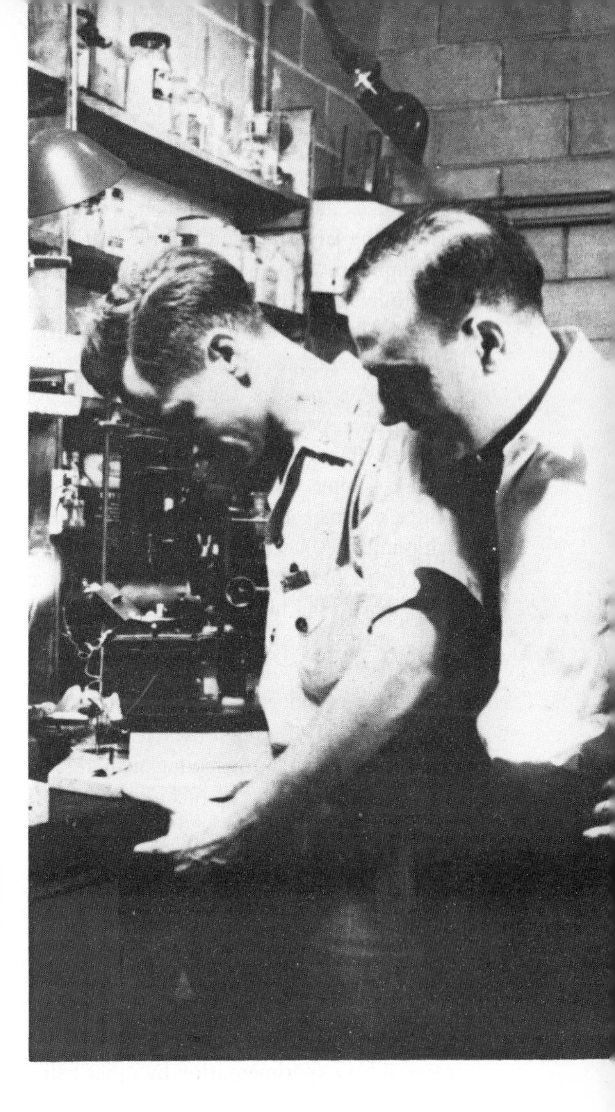

Isadore Perlman and Michael Cefola at work in Room 405 in 1942.

research except that ultramicrochemical techniques would be involved. He strongly stressed the importance of the work without giving specific details.

Since I could not join the group until June, I had ample time to speculate about my role in the laboratories. My first guess was that I would be carrying out analyses of impurities on trace levels. That I was wrong became most obvious when, on a visit to the Met Lab before the official starting date, I saw one of the rooms lined with counting equipment set up by Dr. Spofford English and on the roof the huge evaporating dishes filled with ether extractions of uranyl nitrate solutions.

Since much of my work would center around ultramicrochemical techniques, I was instructed to line up available related equipment. This included microscopes, which were already becoming scarce because of the large demand and the halt of imports from Germany. When equipment trademarked Zeiss later arrived at the laboratory, I was reminded that we were at war with Germany and that U. S. currency would reach Germany through South America.

Initial attempts to restrict unauthorized persons from entering the laboratories on the top floor of the Jones Laboratory were feeble. Later a partition was installed in the corridor leading to Dr. Seaborg's laboratories, and the responsibility for admittance was assigned to some recent high-school graduates and a technical assistant. At times this created amusing situations, such as when Dr. Schlesinger, Chairman of the Chemistry Department at the University of Chicago, knocked on the door and walked in without properly identifying himself. He was immediately stopped by one of the more aggressive young men who asked gruffly, "Where do you think you're going?"

The reply was a meek, "But I am Dr. Schlesinger."

"I don't care if you're Roosevelt. You don't get in here without proper identification," the young man retorted, and, not realizing the stranger's identity, he pushed him outside the entrance.

The difficulty was resolved when Dr. Seaborg identified Dr. Schlesinger.

The equipment from Zeiss and Bausch and Lomb, which included microscopes and micromanipulators, allowed us to carry out reactions on submicro amounts of materials. Even before the isolation of the relatively pure plutonium, many experiments testing the efficiency of different precipitates as carriers and using tracer amounts of plutonium-238 were executed on this scale to conserve the material for experimentation by other members of the group.

Essentially the reactions were performed in capillary cones (similar to centrifuge cones) about 1 millimeter in diameter, 3 to 4 millimeters long and having a capacity of approximately 1 microliter. Because of the cone size, reagents were transferred into them by means of handmade micropipets with tips as small as 0.01 microliter.

Because water evaporated rapidly from such small volumes of solutions, the reaction vessels were placed in a specially designed moist chamber. The entire assembly was then put on a microscope stage where the progress of the reaction could be witnessed.

Immediately following the isolation of microgram amounts of plutonium, this scale of experimentation became extremely useful, especially in the examination of chemical reactions. Conservation of the material was of utmost importance because we needed much information and had so little with which to work.

This milligram-to-submicrogram experimentation substantially reduced the time required to achieve our goal.

GLENN T. SEABORG Mike's remarks reminded me of the curiosity that people had as they joined the project. I used to have the new man, the neophyte, come into my office, and then I had the pleasure of explaining to him that we were working on a new element and watching his consternation and almost unbelievable surprise. In fact, I would often ask him what he thought we were working on, and sometimes the answer would be, "Well, I don't know, but at least I'm sure that it's one of the 92 elements."

I am going to call on Louis Werner next, and that reminds me of the size of the room in which this work was carried on. This was Room 405 of the Jones Laboratory, which is 6 feet wide and 10 feet long. I have taken pleasure on a number of occasions when I describe this work to point out that one of the ultramicrochemists who worked in this room was 6 feet 7 inches tall—that's Louie Werner.

Room 405. . .6 feet wide and 10 feet long. . .

Room 405 in 1942.

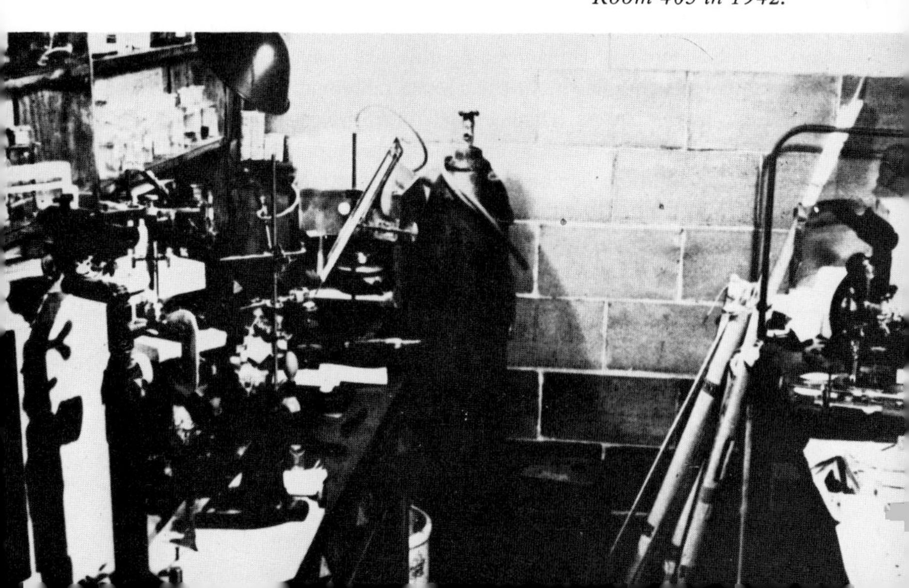

LOUIS B. WERNER Everything did not go smoothly by any means. The most serious problem occurred when I was in the microchemical laboratory and heard a violent clatter in the main part of Jones Laboratory. I went to investigate and found that the centrifuge in which I had placed the world's supply of plutonium had come apart and the solution of plutonium was dripping down through greasy motor bearings onto the floor of the laboratory. That was a black day. Fortunately, by sopping it up with towels and sponges and digesting them, we were able to recover almost all the plutonium. As it turned out, this experience came in quite handy for purifying the first batches of reactor-produced product, which the chemical engineers turned out at Clinton Laboratories in Oak Ridge, Tennessee, and which seemed to contain a little of almost everything the Jones Lab material had picked up.

Once the pure product was isolated, everyone naturally wanted to see what it looked like. However, there was not much to see, and there was some skepticism as to whether there was anything there at all. Rather than tie up the tiny plutonium supply for exhibition, we thought we might engage in a harmless deception and make up a somewhat more impressive solution of simulated plutonium from green ink. Green seemed to be the predominant color of aqueous solutions of plutonium. So the day before an important visitor was scheduled to arrive, we made up a colored solution and set it aside. The next morning, just before our guest arrived, we found to our horror that the green color had turned purple! Unfortunately we did not have the courage of our convictions, because it turned out later that the +3 oxidation state of plutonium *was* purple in color.

There were also educational aspects of our activities during that period. However, we junior scientists were not invited to the weekly technical seminar. Undaunted by this, we organized an independent seminar. A suitable round conference table and meeting room were found in the back of Hanley's Bar on 55th Street, and, with the aid of Hanley's technicians and chemical supplies imported from Milwaukee, these seminars became very popular. The subjects of these conferences tended heavily toward the theory of games and applied statistics. An exercise that was practiced occasionally during

---

The world's supply of plutonium was dripping down onto the floor of the laboratory. . .

these sessions was one that was attributed by his disciples to Professor Charles Coryell. It was called "five card draw, clubs wild."

On a serious note, it was an inspiring experience to be associated with the Plutonium Project and an influence never to be forgotten. It was an opportunity to meet with and be inspired by the current and future famous scientists of the world, many of whom are assembled here today.

Louis B. Werner and Burris B. Cunningham in Room 405 on Aug. 20, 1942.

The peculiar sense of wonder aroused by seeing a new element for the first time. . .

GLENN T. SEABORG It is true that we fibbed a bit and used green dye and even aluminum hydroxide dyed green to represent the plutonium hydroxide. But I remember we mitigated this a little by carefully saying to visitors, at least on occasion, that "this represents a sample of plutonium hydroxide." I do not believe the visitor completely understood the significance of that, but it was not our fault if he thought it actually was a plutonium hydroxide sample.

Now I want to introduce "Iz" Perlman, who played a very important role in the leadership of this plutonium chemistry group. He was an administrator but at the same time a laboratory man, and the fact that, in later stages of the ultramicrochemistry, he actually trained himself as an ultramicrochemist and carried on these intricate experiments is illustrative of his experimental chemical capability.

ISADORE PERLMAN The splendid group we have here today all share the peculiar sense of wonder that was aroused by seeing a new element for the first time. It was the first man-made element seen with the naked eye and an element, indeed, which was to be fateful in the history of humanity. We are commemorating a weighing, and, for those of you who are not chemists, I think the full impact of what it means to weigh something perhaps does not register. Although many elaborate and very important techniques have been evolved to probe into the nature of chemical matter, the touchstone for the evolution of an element's chemistry comes from the preparation of a pure compound, and weighing is a fundamental part of determining that one has a pure compound.

I would like, if I may, to share a little reflective glory in this by telling a story. I do not know what the libel laws are in Illinois, so I will make this short. As already mentioned, there are others who carried out some of the initial steps in the separation of plutonium from rather bulky amounts of material down to where the microchemists could take over. A number of us worked on this. At one stage some of this material fell into my care. As I remember, the material was in a beaker of perhaps a half liter or so when it was put away for the night. The next morning the beaker was broken. A lead

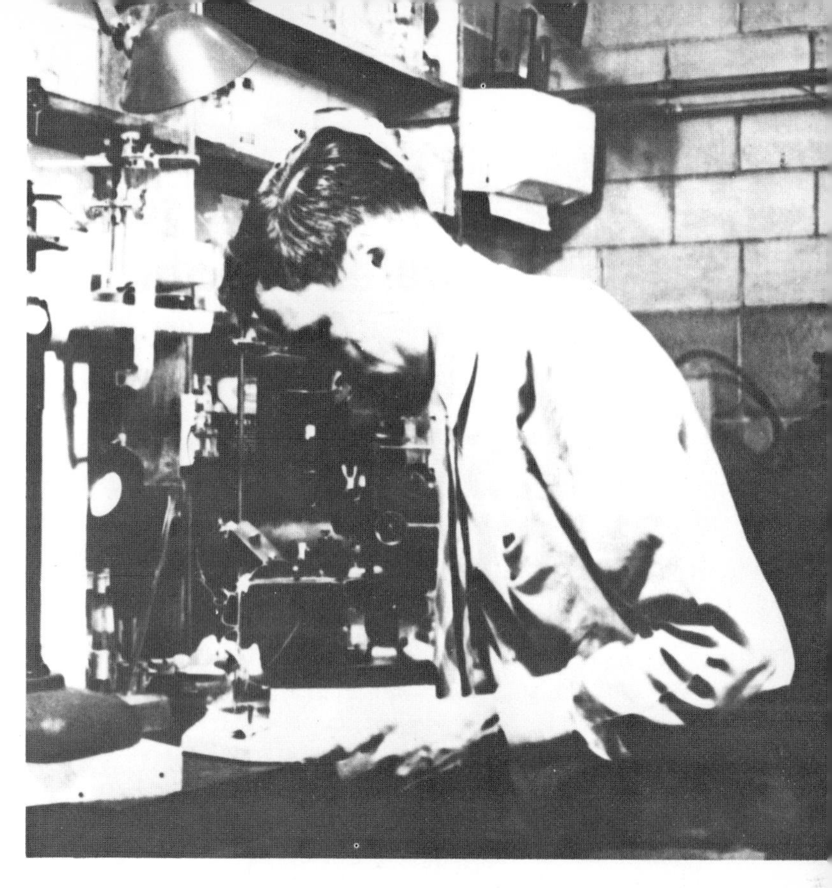

Glenn T. Seaborg in Room 405 on Aug. 20, 1942.

brick had fallen over, and the precious material had spilled. Fortunately, it happened to be sitting on a Sunday edition of the *Chicago Tribune.* A half liter of anything is nothing for this paper to absorb. We proceeded to get the very largest evaporating dish we could find, one larger than a bathtub and smaller than a swimming pool, and we dumped the newspaper in with the idea of subjecting it to what chemists call wet ashing. We digested it with nitric acid and kept this witches brew going for days. That reminds me that this is September 10, and, if it were not for my participation in the isolation of plutonium, this anniversary would probably be celebrated in August.

Well, we finally got all the material in the solution, and I remember vividly that the print still floated. I was very grateful for having that newspaper there, and I cannot avoid reflecting that, among Democrats of left-wing persuasion, I am probably the only one who ever digested an edition of the *Chicago Tribune* so thoroughly. ■

25th Anniversary Celebration of the

FIRST NUCLEAR CHAIN REACTION

In this reactor, under the West Stands of Stagg Field at the University of Chicago, the first self-sustaining atomic chain reaction was achieved on Dec. 2, 1942. The original painting was done in 1968 by Bill Wagoner.

Twenty-fifth Anniversary Celebration of the First Nuclear Reactor, Transmitted over Television from the University of Chicago, Dec. 2, 1967

■ It is a great pleasure for those of us here at Chicago in the United States to join with our friends in Italy to observe this twenty-fifth anniversary of the first nuclear chain reaction.

More than any other person, this anniversary observance honors the great Italian physicist Enrico Fermi, for it was the scientists under his direction at the University of Chicago in 1942 who accomplished the event we are commemorating today—the first controlled release of energy from the nucleus of the atom.

Enrico Fermi's experiment opened up a new world of nuclear matter and energy. Today we are most fortunate that world television, by way of the communications satellite, makes it possible to link the place of discovery, the University of Chicago, with the homeland of the discoverer, Italy.

We are especially fortunate to have with us for this anniversary observance Mrs. Enrico Fermi; many of the members of that group of scientists who worked with her husband in first harnessing the

Laura and Enrico Fermi.

nucleus of the atom; the president of the University of Chicago, Dr. George W. Beadle; and Dr. Herbert L. Anderson, professor of physics at the University of Chicago, who is chairman of the committee that planned this observance.

Now, ladies and gentlemen, the President of the United States.

LYNDON B. JOHNSON, PRESIDENT OF THE UNITED STATES

President Saragat, Mrs. Fermi, Mayor Daley, members of the Fermi team, Dr. Seaborg, and distinguished guests:

I believe history will record that on this day 25 years ago mankind reached the turning point of his destiny.

The Book of Genesis tells us that, in the beginning, God directed man to "Be fruitful, and multiply, and replenish the earth, and subdue it. . . ."

Scientists who were present on Dec. 2, 1942, when the first reactor operated celebrate the twenty-fifth anniversary of the first self-sustaining atomic chain reaction. Left to right, front row: Leo Seren, H. M. Barton, Gerard S. Paulicki, August C. Knuth, George Miller, R. J. Fox, Wilcox P. Overbeck, Crawford Greenewalt, and Walter Zinn; middle row: Herbert Anderson, Warren E. Nyer, Harold M. Agnew, Albert Wattenberg, William H. Hinch, and Stewart A. Fox; back row: W. R. Kanne, Thomas Brill, Norman Hilberry, Herbert E. Kubitshek, Volney C. Wilson, George D. Monk, Harold V. Lichtenberger, M. H. Wilkening, E. O. Wollan, Leon Petry, Theodore Petry, William J. Sturm, and Robert E. Johnson.

Site of the first reactor, the West Stands of Stagg Field, University of Chicago.

Fourth anniversary of the first self-sustaining atomic chain reaction. Scientists meet on Dec. 2, 1946, to celebrate their success. Left to right, back row: Norman Hilberry, Samuel Allison, Thomas Brill, Robert G. Nobles, Warren Nyer, and Marvin Wilkening; middle row: Harold Agnew, William Sturm, Harold Lichtenberger, Leona W. Marshall, and Leo Szilard; front row: Enrico Fermi, Walter H. Zinn, Albert Wattenberg, and Herbert L. Anderson.

The modern Italian navigator was a great man of science. . .

But only in our lifetime have we acquired the ultimate power to fulfill all of that command. Throughout history, man has struggled to find enough power—to find enough energy—to do his work in the world. He domesticated animals; he sold his brother into slavery; and he enslaved himself to the machine—all in a desperate search for energy.

In Chicago 25 years ago, Enrico Fermi and his fellow scientists, in a single stroke, increased man's available energy more than a thousandfold. They placed in our hands the power of the universe itself.

Nothing could have been more appropriate than the words used by Dr. Arthur Compton to describe what happened on that day:

"The Italian navigator has just landed in the new world."

This modern Italian navigator was a great man of science. But he was also something more. He was one of millions who, in the long history of the world, have been compelled to leave a beloved native land to escape the forces of tyranny. And, like millions before him, Enrico Fermi found here a new home among free men in a new world. His life and his career have a very special meaning to all who love freedom.

There are today millions of young Americans with an Italian heritage who feel a deep personal pride in Enrico Fermi. America was born out of the voyages of a great Italian navigator. In a time of greatest danger, another—equally willing to pursue his dream beyond existing charts—took us again into a new epoch.

Today we commemorate our debt to him. And in doing so we also honor the historic bond between the Old World and the New World.

In a short time we will be dedicating in the great state of Illinois a new national accelerator laboratory. This laboratory, with its 200 billion electron volt accelerator, will maintain our country's position in the forefront of nuclear research.

I suggest that we dedicate this great new laboratory to the memory of the modern-day "Italian navigator."

In so honoring Enrico Fermi, we will also honor the immeasurable contributions that have been made over the centuries by the people of Italy to the people of the United States.

Enrico Fermi when he was a Professor of Physics at the University of Chicago.

Giant nuclear reactors, direct descendants of Fermi's first pile, are now producing millions of kilowatts of power for peaceful purposes. . .

The first reactor reassembled to the 19th layer.

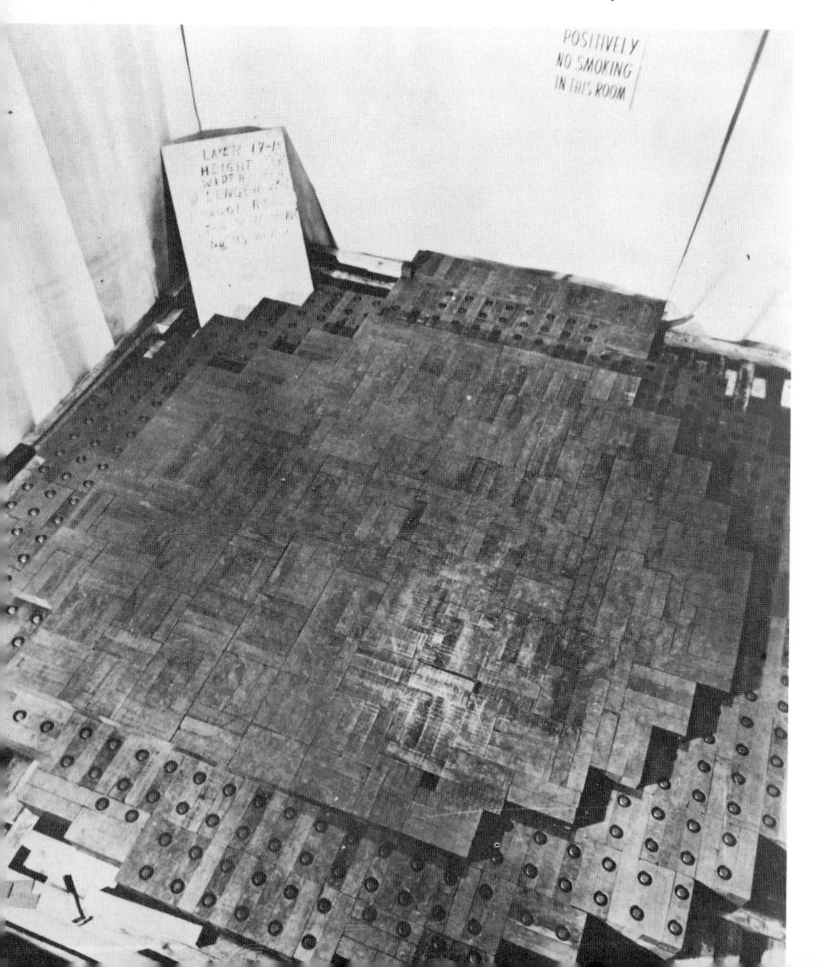

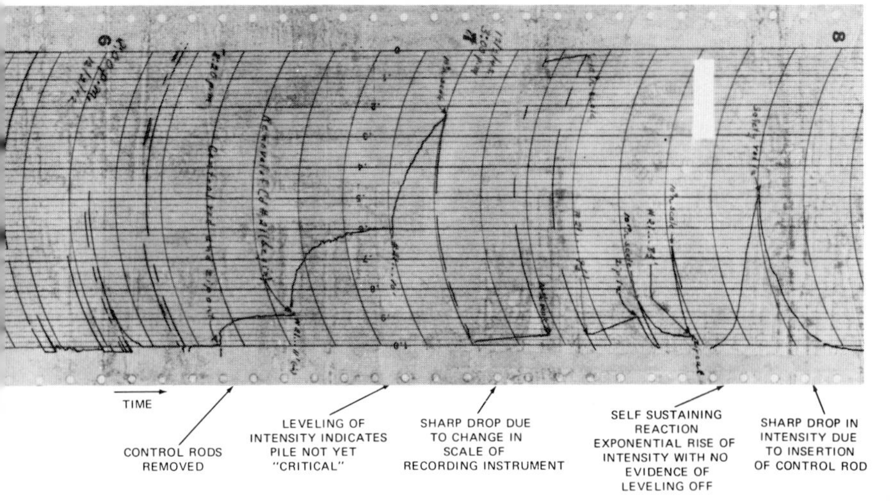

The "birth certificate" of the Atomic Age. The graph indicates the neutron intensity as recorded by a galvanometer during operation of the first pile.

'his drawing was in an appli-ation to the U. S. Patent)ffice filed by Enrico Fermi nd Leo Szilard in December 944, for a patent on the first uclear reactor.

Much has already happened in that new world that began just 25 years ago.

Giant nuclear reactors, direct descendents of Fermi's first atomic pile, are now producing millions of kilowatts of power for peaceful purposes. Other reactors are powering nuclear submarines under the seas. They are our first line of defense against tyranny, whatever its contemporary doctrine or disguise, which Enrico Fermi dedicated himself to resist.

But it is really the peaceful uses of atomic energy about which Fermi would have wished us to speak—and there are many peaceful uses.

When I became President, nuclear energy was generating about 1 million kilowatts of electric power in the United States.

Today, the atom is giving us more than 2,800,000 kilowatts—almost three times as much. And more than 70 additional nuclear powerplants are already planned or are now under construction.

This will equal about 20% of the whole electric generating capacity in the United States today. It is enough to meet the total requirements of 45 million people. All this from what was 25 years ago—before the success of Fermi's experiment—only a scientist's dream.

The dream has been realized. By learning the secret of the atom, we have for the first time in history given mankind all the energy that it can possibly use.

It took the genius of countless generations of dedicated scientists to find the secret. It remains for us to use that secret wisely.

What began as the most terrible instrument of war that man has ever seen can become the key to a golden age of mankind. But this will not happen unless we make it happen.

We cannot forget that another, darker future also opened on this day 25 years ago.

The power to achieve the promise of Genesis is also the power to fulfill the prophecy of Armageddon. We can either remake life on earth or we can end it forever.

Let me be specific.

If Enrico Fermi's reactor had operated 10,000 years, it would not

Fermi closed his slide rule. "The reaction is self-sustaining," he announced quietly.

The "Council Tree" in front of Eckhart Hall on the University of Chicago campus beneath which scientists held a highly secret discussion in April 1942 that was vital to the success of the first reactor. They met here so they could talk freely without being overheard.

have produced enough plutonium for one atomic bomb.

Today a single reactor can, while generating electricity, produce enough plutonium to make dozens of bombs every year. And scores of these reactors are now being built—and they are being built all over the world.

Their purpose is peaceful. Yet the fact remains that the secret diversion of even a small part of the plutonium created could soon give every nation—every nation—the power to destroy civilization, if not life on this earth.

We just cannot permit this to happen. Nor can mankind be denied the unlimited benefits of the peaceful atom. We must, some way, somehow, find a way to remove the threat while preserving the promise.

The American people have made their own desires crystal clear when their representatives in the United States Senate voted unanimously to support an effective nonproliferation treaty for nuclear weapons.

We are now engaged in a major effort to achieve such a treaty, in a form acceptable to all nations.

We are trying hard to ensure that the peaceful benefits of the atom will be shared by all mankind without increasing, at the same time, the threat of nuclear destruction.

We do not believe that the safeguards we propose in that treaty

The Congressional Medal of Merit that Gen. Leslie R. Groves pinned on Enrico Fermi on Mar. 19, 1946, read in part, "As the pioneer who was the first man in all the world to achieve nuclear chain reaction ---."

will interfere with the peaceful activities of any country.

I want to make it very clear to all the world that we in the United States are not asking any country to accept safeguards that we are unwilling to accept ourselves.

So I am today announcing that when such safeguards are applied under the treaty the United States will permit the International Atomic Energy Agency to apply its safeguards to all nuclear activities in the United States—excluding only those with direct national security significance.

Under this offer the agency will then be able to inspect a broad range of United States nuclear activities, both governmental and private, including the fuel in nuclear powered reactors owned by utilities for generating electricity and the fabrication and the chemical reprocessing of such fuel.

This pledge maintains the consistent policy of the United States since the very beginning of the nuclear age.

It was just 14 years ago that a President of the United States appeared before the General Assembly of the United Nations to urge the peaceful use of the atom. President Dwight D. Eisenhower said on that occasion:

> . . .the United States pledges . . . before the world . . . its determination to help solve the fearful atomic dilemma—to devote its entire heart and mind to find the way by which the miraculous inventiveness of man shall not be dedicated to his death, but consecrated to his life.

We renew that pledge today. We reaffirm our determination to dedicate the miraculous power of the atom, not to death, but to life.

We invite the world's nations to join with us.

Let us use this historic anniversary to deepen and to reaffirm the search for peace.

Let us so conduct ourselves that future generations will look back upon Dec. 2, 1942, not as the origin of sorrow and despair, but as the beginning of the brightest and the most inspiring chapter in the long history of man. ■

At the fifth anniversary of the first self-sustaining atomic chain reaction on Dec. 2, 1947. The plaque on the wall of the West Stands, Stagg Field, was unveiled then.

THE DISCOVERY OF URANIUM-233

Plaque reads: "U-233—The fissionability of U-233 was demonstrated in this room by G. T. Seaborg, John W. Gofman, and Raymond W. Stoughton on February 2, 1942."

John W. Gofman, Glenn T. Seaborg, and Raymond W. Stoughton (left to right) on the 25th anniversary of the discovery of the fissionable nature of uranium-233.

To the Regents of the University of California, Berkeley, California, Jan. 19, 1967

■ There is another man-made nuclear fuel that may be the equal of plutonium-239 in energy potential and in ultimate importance to man. This fuel is an isotope of uranium, uranium-233, which should not be confused with the familiar uranium-235, about which I shall say more later.

Just as the fissionable isotope plutonium-239 is the key to the unlocking of the vast amount of energy stored in the abundant but nonfissionable isotope of uranium, uranium-238, occurring in nature, so the fissionable isotope uranium-233 is the key to the unlocking of the enormous energy stored in the abundant and again nonfissionable isotope of thorium, thorium-232, found in nature. Uranium-233 and plutonium-239 taken singly or in combination provide man with an almost infinite source of energy sufficient for centuries to come.

It is not generally recognized that this second nuclear fuel, uranium-233, like plutonium-239, was first created by use of the Berkeley 60-inch cyclotron and was first identified and then found to be a potential nuclear fuel in the same suite of laboratories on the third floor of Gilman Hall on the Berkeley campus. The cast of scientific characters who brought uranium-233 into the world was somewhat different from that engaged in creating plutonium. The labors in search of the two man-made isotopes paralleled each other. Thus two weeks from today will be the twenty-fifth anniversary of the evening of Feb. 2, 1942, when John W. Gofman, Raymond W. Stoughton, and I were able to say that we had created and identified a second major source of nuclear energy. Although the chemical separation of uranium-233 was carried on in Room 307, the same room in which plutonium was discovered, the important verification of the fissionability of uranium-233 with slow neutrons (in other words, its ability to sustain a fission chain reaction and thus its capability of fueling the fires of a nuclear reactor) was carried on in Room 303 Gilman.

I am pleased that Drs. Gofman and Stoughton are here with me today. Both have had distinguished careers since 1942. Dr. Gofman went on to become a physician and biomedical researcher. At the

---

Uranium-233, like plutonium-239, was first created by use of the Berkeley 60-inch cyclotron. . .

The 60-inch cyclotron in August 1939.

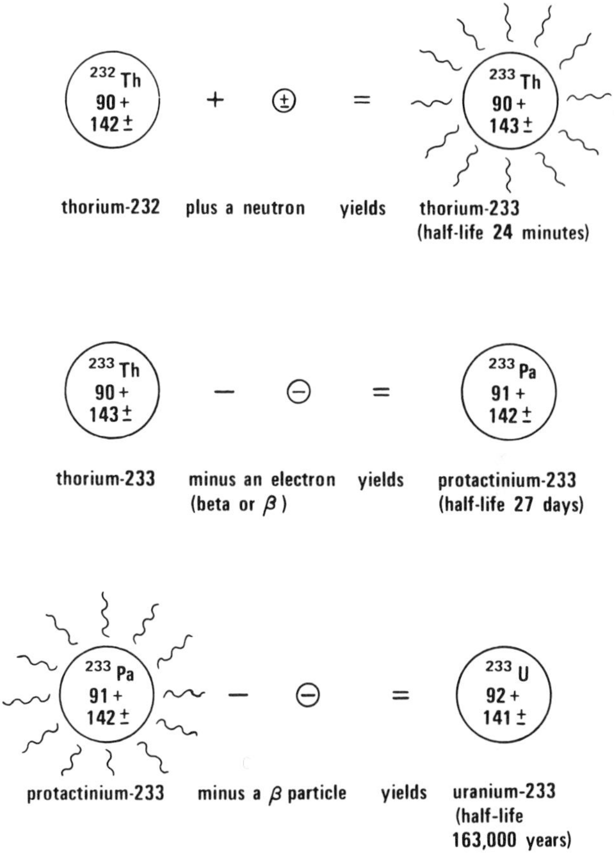

^{232}Th 90+ 142±	+	±	=	^{233}Th 90+ 143±
thorium-232	plus a neutron		yields	thorium-233 (half-life 24 minutes)

^{233}Th 90+ 143±	−	⊖	=	^{233}Pa 91+ 142±
thorium-233	minus an electron (beta or β)		yields	protactinium-233 (half-life 27 days)

^{233}Pa 91+ 142±	−	⊖	=	^{233}U 92+ 141±
protactinium-233	minus a β particle		yields	uranium-233 (half-life 163,000 years)

Donner Laboratory he distinguished himself in the applications of radioisotopes in medicine, in research on the role of fatty molecules in arteriosclerosis, and in other studies. Today he is Professor of Medical Physics at Berkeley and Associate Director of the Lawrence Radiation Laboratory, Livermore, for biomedicine. There he is leading an important effort to assess the hazards of man-made radiation and to provide the means of protecting man from those hazards in a future in which radioactivity will be generated in increasing quantities. Dr. Stoughton continued to work on the development of the potential of uranium-233 as a nuclear fuel, first at the Metallurgical Laboratory in Chicago and then at the Oak Ridge National Laboratory in Tennessee. Since 1943 he has been a staff member at the Oak Ridge National Laboratory, where his work on the chemical properties of uranium in aqueous solution and his other effective contributions, especially in high-temperature aqueous solutions, have established for him an outstanding reputation as an inorganic chemist.

As I thought about President Kerr's kind invitation to me to say a few words to you about the discovery of uranium-233, I became involved in a bit of amusing fantasy. What would have been the reaction of an obscure young chemistry professor named Seaborg if, through the medium of some Wellsian time machine, the future as it has evolved in this last quarter of a century had been revealed to him 25 years ago? I assure you that the young nuclear chemist would have greeted the revelations with disbelief if not derision. And a good thing too. For had he believed and attempted to make believers of others, he surely would have spent a good part of the last 25 years in some institution that accommodates people with delusions.

I think I can illustrate this by briefly contrasting the situation a quarter of a century ago, and the rather unformed prospects of that time, with today's realities in nuclear energy. We started the search for plutonium and uranium-233 in 1940, at the end of a decade of expansion of knowledge of the atomic nucleus fostered by a rich collaboration of European and American scientists. Much of this development was made possible here at Berkeley with the cyclotrons of the late Ernest O. Lawrence, whose genius and inspiration were so important in making the Nuclear Age a reality.

In the fall of 1940, we were still thinking and working primarily in the traditional academic manner. There was no government support for our research. Fortunately California believed in higher education and in the importance of graduate study and research in the university. Our faculty salaries were paid by the university—I remember that my salary was $200 a month in the fall of 1940—a sum that, in view of my experiences in the depression, seemed like incredible wealth. We had some basic facilities, small funds for equipment, and, when we were lucky, modest grants from generous private donors and foundations.

The basic research on the atomic nucleus carried on throughout the world had resulted in the surprising discovery of nuclear fission in 1939 by two German scientists, Hahn and Strassmann. A new powerful potential source of energy emerged. But most scientists were slow to think seriously of the new energy source as one that could be harnessed in a short period of time. For example, it was discovered quickly that the natural isotope of uranium that fissioned with slow neutrons was uranium-235, which comprises less than 1% of natural uranium, and the separation of this isotope in quantity from natural uranium was initially considered to be a staggering and,

Uranium-233 and plutonium-239 provide man with an almost infinite source of energy. . .

Jubilation and ceremony marked the startup in October 1968 of the Oak Ridge National Laboratory Molten Salt Reactor, which uses uranium-233. Oak Ridge National Laboratory Director Alvin Weinberg points out a detail of the control panel to Glenn T. Seaborg, codiscoverer of uranium-233.

some believed, an impossible prospect. Two other avenues deserved some exploration. It might be possible, by adding a neutron to the plentiful uranium-238, to manufacture an isotope of element 94 with a mass number of 239, now known as plutonium-239. Such an atom, which does not exist in nature, might undergo fission with slow neutrons, as uranium-235 does, and, being a different chemical element, might be produced in pure form more easily than uranium-235 could be separated from its sister isotopes of uranium. Similiarly, it was thought possible, by adding a neutron to the abundant thorium-232, to create another fissionable isotope of uranium, uranium-233, which does not occur in nature. This new uranium isotope could be produced in the absence of other naturally occurring uranium isotopes and be simply separated by known chemical means from its parent, thorium.

Some European refugees and American scientists feared that Nazi Germany might somehow make a super weapon with nuclear energy, in which case Hitler would have Britain and the United States at his mercy. But, in the fall of 1940, the possibility that nuclear energy might become a factor in World War II was not really appreciated. The ideas for harnessing nuclear energy at that time were pretty much in the blue-sky category. In addition to the difficulties in separating uranium-235, the species plutonium-239 and uranium-233 had never been made. Nor was there assurance that they could be made or that, if made, they would undergo fission with slow neutrons. Moreover, the government did not provide funds for the search for plutonium until the summer of 1941 and for uranium-233 until the following fall. I do not mean to imply that we did not take our work seriously. On the contrary, on the chance that something practical might emerge, we academic scientists voluntarily imposed our own system of secrecy over the work even though the government had not become involved. Certainly, however, what was in store for us, for the country, and for mankind was by no means clear.

Using resources provided by the University of California and private donors, we began our two searches pretty much in the same way we had done earlier research. Dr. Edwin M. McMillan, now Director of the Lawrence Radiation Laboratory, and Dr. Philip H. Abelson, now Director of the Geophysical Laboratory at the Carnegie Institution in Washington, D. C., had discovered element 93, neptunium, in the spring of 1940. In further work McMillan obtained data suggestive of, but not conclusive for, the existence of element 94. When McMillan was called to M.I.T. to do research on radar, I asked him, in the academic tradition, if I might continue his line of investigation. He approved. Subsequently, as I related last year, the late Joseph W. Kennedy, Arthur C. Wahl, and I proved the existence of plutonium beyond doubt on Feb. 23–24, 1941. With the collaboration of Dr. Emilio Segrè of the Radiation Laboratory, we went on to demonstrate the fissionability of plutonium-239 with slow neutrons about a month later.

Meanwhile, I was looking for a good graduate student to begin work on the potential isotope uranium-233. Jack Gofman showed up in Berkeley for the fall semester of 1940 after a postgraduate year as a medical student at Western Reserve University in Cleveland, Ohio. He had decided to make a career of medical research with heavy emphasis on chemistry and had come to Berkeley to gain an extensive background in chemistry. His first stop at Berkeley was a talk with G. N. Lewis, the "father of chemistry" here. Lewis advised Gofman to "go shopping" for a professor and to get to work on his research in a couple of weeks, a suggestion that appalled the young man. Fortunately Gofman and I chose each other, and within two weeks he was indeed getting his feet wet in the laboratory. Jack tells me that I told him of uranium-233 that "it's not a bad problem for a thesis."

That fall we cleared away some preliminary questions about nuclear reactions leading to the formation of uranium-233. In early 1941 we made our first attempt to produce uranium-233 by bombarding thorium in the neutron beam of the 60-inch cyclotron. This produced little because the 10-gram target was too small and the hour-long bombardment was too short. We then prepared a bigger (kilogram) sample of thorium nitrate and irradiated it in the neutron beam of the cyclotron for weeks. Jack then spent most of the spring working up the sample chemically in Room 307 Gilman.

The suspected nuclear reaction was as follows: Thorium-232, upon capturing a neutron, would be converted into thorium-233, which was known to have a short half-life (about 24 minutes). Then the thorium-233 would emit a beta particle and be transformed into protactinium-233, with a half-life of 27 days. Finally, the protactinium-233, upon emitting a beta particle, would turn into uranium-233. According to the theory, uranium-233 should decay by emitting an alpha particle—a rather heavy particle.

Our first problem was to separate protactinium-233 chemically from the thorium nitrate. The protactinium-233 would turn into uranium-233 rather rapidly, and we could watch for some alpha particles growing into the sample; this would be indicative of the presence of uranium-233 if our theory was correct. As you can see from all this, a number of different kinds of radiation were involved,

Our work was interrupted by an incident that seems almost incredible. . .

---

and one of the problems was to be able to identify the alpha particles when and if they emerged from the sample. The late Dr. Joseph W. Kennedy, a genius at developing counting techniques, developed the instrumentation for detecting alpha particles in the presence of beta-particle radiation.

Our array of counters was in Room 303 Gilman. This small room also served as an office for Joe Kennedy and me. I assure you it was crowded, with our desks, file cabinets, workbench, counters, electrometers, and eventually alpha counters and fission chambers. But we really did not know it was crowded at the time. Compared to certain of the facilities I had used in some of my initial work in nuclear chemistry—including an abandoned building and what I think was a janitor's washroom—it was a luxury.

By the end of the spring semester of 1941, we had observed alpha particles emerging from our sample and were confident that we had discovered the long-lived alpha-particle emitter uranium-233. But we needed to produce a much bigger sample to establish the fission properties of uranium-233.

At this point our work was interrupted by an incident that, in the light of today's level of scientific effort, seems almost incredible. Gofman, like virtually all students in those days and a very large percentage today, was poor. During the school year he had supported himself on the small pay of a teaching assistant. But when the semester ended, the job also ended. There were no summer jobs to be had. I tried without success to get money to pay Jack for working through the summer in the laboratory. He tried to get an off-campus job so he could remain in Berkeley and work part-time in the laboratory without pay. But all our efforts were fruitless. So Jack went back to Cleveland to live with his family until fall, when the job as teaching assistant would again be available. The plutonium problem was assuming an urgent phase, and I could not carry on the further investigation of uranium-233 alone. Thus the search for this fantastic new source of energy ceased during the summer of 1941.

By fall, when Jack returned to his teaching assistant's job and to the work on uranium-233, the United States government had provided a modest grant for this research. Although Jack continued to be supported solely by his university salary as a teaching assistant, the funds included $3000 to enable me to hire a Ph.D. research chemist. I received authorization to do this in response to my letter to Washington of July 10, 1941, which included the following entreaty:

> In case it is decided that a contract for these projects, with a chemist assistant, is to be assigned to me, could you authorize me to hire a chemist to start to work as soon as possible without waiting for the official completion of the contract negotiations? Good unemployed chemists are becoming increasingly difficult to find, and I know of one who will be available provided I can give him some definite information soon.

The government funds enabled me to invite Ray Stoughton to join the project. Ray was a valuable member of our team, and his shouldering of a good part of the heavy burden of chemical processing was an essential ingredient in our final success.

As soon as possible we put about 3 kilograms of thorium nitrate in the cyclotron's neutron beam. By Feb. 2, 1942, a 4-microgram sample of uranium-233 had been isolated chemically in Room 307 Gilman; this was a large sample by the standards of the day. The sample was taken to the clutter of Room 303 where we could easily count the telltale alpha particles characteristic of uranium-233 decay. The fissionability of uranium-233 with slow neutrons was determined that same night and early the following morning. In a later experiment the half-life of this isotope was established as about 100,000 years.

Thus there were, and are now, three potential sources of nuclear energy. One was uranium-235, the scarce isotope of natural uranium, which comprises less than 1% of that element in nature. The other two were man-made: plutonium-239, which could be manufactured in a nuclear reactor from plentiful nonfissionable uranium-238, and uranium-233, which could be made in a nuclear reactor from plentiful nonfissionable thorium-232.

Even with nuclear development put on a crash basis, one would have been hard put to visualize the rapidity and scope of its development. Twenty-five years ago I could not have imagined the existence of the huge and diversified nuclear energy enterprise of the federal government or that I would have a considerable responsibility for making nuclear energy a practical and economic reality.

Let us look briefly at the present status of nuclear energy and the future roles of these three nuclear energy sources, including the two man-made ones discovered at the University of California. The original fissionable material, and the only one occurring in nature, is the rather scarce uranium-235. In view of the huge amounts of energy released in fission, uranium-235 seemed to be a vast energy resource despite its relative scarcity. And so it is, in the sense that it provides a resource of energy exceeding that available from fossil fuels. However, used as the sole source of nuclear energy, and keeping in mind a long time scale, uranium-235, like fossil fuels, could be spent fairly quickly.

But, with the indirect burning of abundant uranium-238 and thorium-232 as nuclear fuels through the intermediate use of fissionable plutonium-239 and uranium-233, nuclear energy becomes a vast and, for all practical purposes, a virtually unlimited energy resource. There is over 100 times as much ordinary uranium (uranium-238) that can be converted into plutonium-239 as there is uranium-235. And the thorium (thorium-232) resources are about equal to, or greater than, those of ordinary uranium.

The discovery of these two man-made nuclear fuels here in Berkeley within one year constitutes, as we contemplate the future of nuclear electrical power generation, one of the most impressive payoffs on basic research in history. These discoveries, like nuclear fission itself, were the result of a decade of rapid expansion of knowledge on what appeared to be an abstruse and apparently impractical subject, namely, What makes the atomic nucleus tick?

I have described a number of aspects of this payoff, in particular

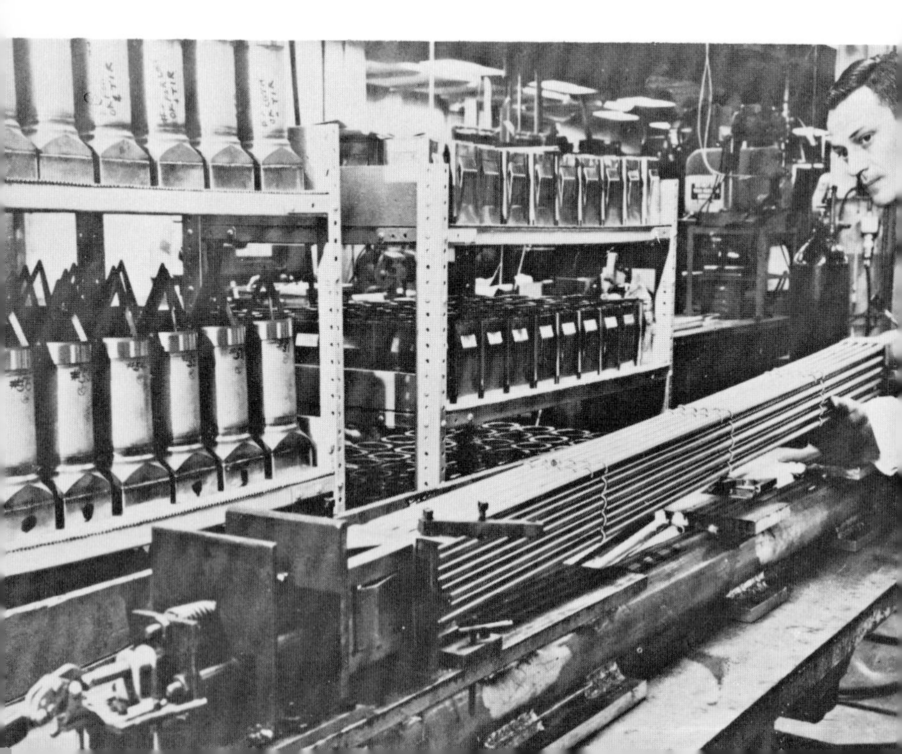

the assurance of abundant energy for civilization for hundreds of years and the significance of abundant energy to the world of the future. There is another way to express the payoff, and that is in terms of dollars. When we compare the amount of energy derivable from U. S. uranium reserves through the uranium–plutonium-239 cycle with the cost of energy from fossil fuels today, the value of the potential nuclear energy comes to something around 50 quadrillion dollars. The value of uranium-233 derivable from thorium is estimated to be of a similar order of magnitude.

We are all aware, of course, that the release of nuclear energy was pressed initially as a means of preserving freedom against tyranny. In the future uranium-233 and plutonium-239 will be important instruments for man's continued security and material abundance. And men can have cause to celebrate the events that took place under the eaves of Gilman Hall on the Berkeley campus of the University of California a quarter of a century ago. ■

Fuel bundles for the Elk River boiling-water reactor used urania–thoria pellets sealed inside stainless-steel tubes.

ELEMENTS 95 and 96

25 Years Ago

Americium

First sample of an americium compound, isolated in January 1946. Eye of needle shows degree of magnification.

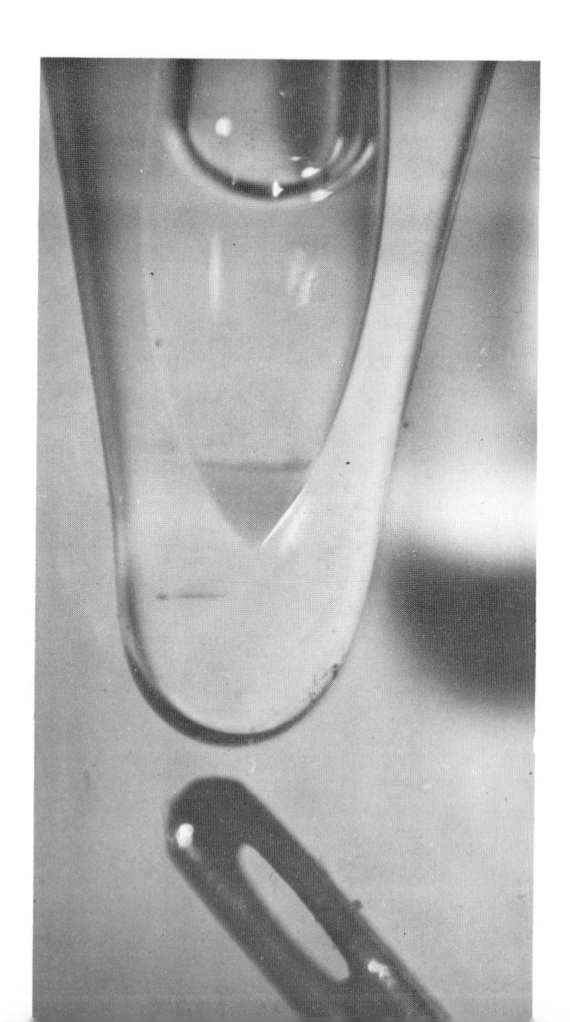

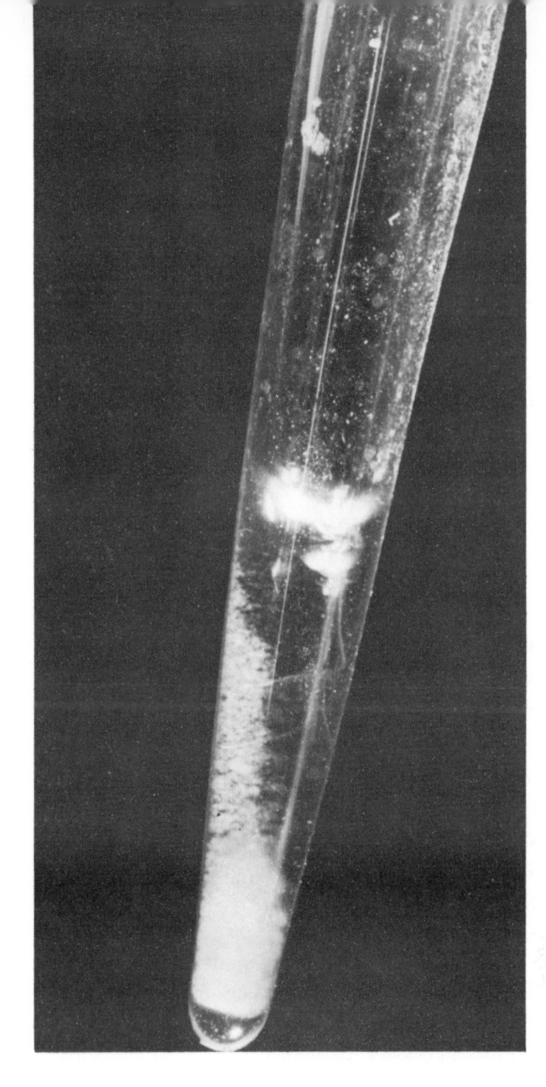

Curium

First sample of a curium compound, isolated in 1947, distributed as a precipitate in the capillary tube.

At the Twenty-fifth Anniversary Observance of the Discovery of Americium and Curium, Robert A. Welch Foundation Conference. XIII: Transuranium Elements—The Mendeleev Centennial, Houston, Texas, Nov. 17, 1969

■ We are gathered here to observe the twenty-fifth anniversary of the discovery of two very interesting synthetic elements—the transuranium elements with the atomic numbers 95 and 96. The discovery experiments for these elements, which were given the names americium and curium, were performed at the University of Chicago in the wartime Metallurgical Laboratory in the New Chemistry Building.

I am particularly pleased that my coworkers in these experiments,

The discovery experiments took place during 1944 and early 1945. . .

Albert Ghiorso, Ralph A. James, and Leon O. (Tom) Morgan, can be with us and participate in this program. I am also glad to see so many others who were members of the Metallurgical Laboratory when this work took place or who have a special interest in these elements.

The discovery experiments took place during 1944 and early 1945. In the course of my remarks, I shall attempt to trace for you a sort of blow-by-blow account, including the dates when the critical observations were made. I found the preparation of this talk a most fascinating endeavor, steeped in nostalgia as I searched the deep recesses of my memory, picked the brains of my colleagues, and pored over our notebooks of that era of 25 years ago. This is in itself a moving experience, bringing back as it does the many emotional reactions—both heartaches and triumphs—of that time. The understanding and interpretation of the results took a longer period of time and had more of the ingredients of a detective story than was the case for the other elements in whose discovery I had the privilege to participate.

A number of factors contributed to the special status that this research holds in my memory. The experimental tools available at that time were crude in relation to the deductions that had to be made. And a real breakthrough in thinking had to be made to devise the chemical procedures necessary to identify the new elements. I believe that I derived more personal satisfaction from these experiments than from those concerned with the discovery of the other transuranium elements.

New Chemistry, the building where this work was done, stood on the east side of Ingleside Avenue extending south from 56th Street. It was hastily constructed during the fall of 1942 when it became apparent that the quarters on the fourth (top) floor of George Herbert Jones Laboratory were grossly inadequate for our group of chemists working as part of the Metallurgical Laboratory. I was responsible for the chemists working on the chemical processes to be used in the extraction of the plutonium from neutron-irradiated uranium and on the basic chemistry and purification of plutonium, and I, of course, was devoting my full time and energy to this task.

By December of 1943, however, so much progress had been made on these problems that I felt I could devote part of my efforts to the synthesis and identification of the transplutonium elements with the atomic numbers 95 and 96. I asked Ralph James, a young chemist from Berkeley who had been especially proficient in the investigations of the radiochemistry of plutonium, to devote himself to this problem.

We began by attempting to produce isotopes of element 95 through the bombardment of plutonium-239 with deuterons. The target material, plutonium-239, was then just becoming available in milligram amounts as a result of its production in the uranium–graphite reactor at the Clinton Laboratories in Tennessee.

Ralph and I went to St. Louis at the end of January 1944, and with the help of Harry Fulbright, we bombarded 0.1- and 1-milligram samples of plutonium-239 with the deuterons furnished by the cyclotron at Washington University. Measurements were made on the bombarded plutonium-239, immediately at St. Louis and later in Chicago, without any attempt to chemically separate the products. Alpha-particle-range measurements were made to try to detect the presence of alpha-particle emitters with higher or lower energy than that of plutonium-239. These crude experiments gave negative results.

The 1 milligram of irradiated plutonium-239 and two other larger samples that were bombarded with deuterons during the following months (one in the cyclotron at Washington University and the other in the 60-inch cyclotron in the Radiation Laboratory of the University of California, Berkeley) were subjected to a number of

Leon O. Morgan, Chicago, 1944.

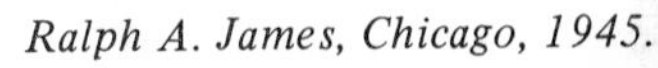

Ralph A. James, Chicago, 1945.

We irradiated samples of plutonium-239 to look for the production of plutonium-240. . .

---

chemical procedures. Chemical fractions were isolated and their radiations examined on the basis of several hypotheses concerning the chemical properties of element 95. These hypotheses included various assumptions concerning the solubility properties of the compounds of element 95 and its potential for oxidation from a lower state in which its fluoride is insoluble to an upper state in which its fluoride is soluble—including the assumption that it could not be so oxidized at all. (Each of the elements immediately preceding element 95—uranium, neptunium, and plutonium—has III and IV oxidation states that have insoluble fluorides and a VI state that has a soluble fluoride. As we go from uranium to neptunium to plutonium, increasingly strong oxidizing agents are required to attain the VI state.) Unique alpha particles with energies different than those of plutonium-239 were looked for, but in no case were they found. In retrospect, this is not surprising; we now know that the isotopes that would have been produced, such as 95-240 and 95-239, decay overwhelmingly by electron capture and to such a small extent by alpha-particle emission that this could not have been detected with the techniques available. These techniques did not permit detection of electron-capture decay in the presence of the tremendous amount of rare-earth fission-product (beta and gamma) activity.

Also, beginning in January 1944 in parallel experiments, we used the reactor at the Clinton Laboratories in Tennessee to irradiate samples of plutonium-239 with neutrons. Although it was thought at that time to be unlikely, we wanted to look for the production of plutonium-240. If such a (n,γ) reaction could occur to an appreciable extent in competition with fission, then successive (n,γ) reactions might occur, leading to a beta-particle-emitting plutonium isotope and hence to an element 95 daughter. Again, in retrospect, we now know that the similarity of the alpha-particle-decay properties of plutonium-240 and plutonium-239 made impossible the detection of the small concentration of plutonium-240 present in those neutron-irradiated samples.

An important factor in the measurement of the radiation from the isotopes of elements 95 and 96 being sought was the

Albert Ghiorso and Arthur Jaffey in the new Chemistry Building, Metallurgical Laboratory, University of Chicago, in 1946.

determination of the energy of alpha particles by absorption or range measurements. The early absorption measurements I shall report here were performed with thin mica absorbers and are translated into equivalent range in terms of centimeters of air at standard conditions. During this period Albert Ghiorso, the group leader of our Instruments and Physical Measurements Group, began to give more and more of his time to this problem. As the work progressed he devised instrumentation and techniques of increasing complexity as required. Other members of his group, particularly Arthur H. Jaffey, also gave us very valuable help.

The experiments performed in the first half of 1944 gave negative results but provided much valuable experience. The first breakthrough came in July. The first bombardment of plutonium-239 with helium ions (32 MeV) took place in the Berkeley 60-inch cyclotron July 8–10. By this time the actinide concept had crystallized in my mind to the extent that we decided to proceed solely on the assumption that elements 95 and 96 could not be oxidized in aqueous acidic solution to soluble fluoride states (i.e., they should exhibit only the III or IV oxidation states having insoluble fluorides). The sample was sent to Chicago by air, and the chemical procedures were started on July 12. A 2.2-milligram portion of the target plutonium-239 was oxidized in aqueous solution to the plutonium (VI) oxidation state (whose fluoride is soluble), and lanthanum fluoride was precipitated. This precipitate, presumably containing the insoluble fluoride of an element-95 or -96 isotope and a small fraction of the plutonium (nearly all of which remained in solution in its soluble oxidized form), was then dissolved. Most of the small amount of remaining plutonium was oxidized, and lanthanum fluoride was again precipitated to carry the element 95 or 96. The cycle was then repeated a third time. This final sample (labelled 49αA – #9) contained only about 0.004% of the original 2.2 milligrams of plutonium-239, i.e., about 0.09 microgram or about 12,000 alpha-particle disintegrations per minute.

On July 14, 15, and 16, careful absorption measurements in mica sheets were made on this sample. A plot of these data revealed distinctly the presence of alpha particles of longer range than those of plutonium-239. Perhaps I can best summarize the significance of this observation by quoting here the entry we made in our notebook (No. 221B) on July 14, 1944:

> Plotting and comparing this data shows that a 94^{239} sample 20% larger than 49αA – #9 falls to zero c/m much faster than does 49αA – #9 itself. This definitely shows the presence of a long-range alpha emitting isotope in sample 49αA – #9. This is undoubtedly due to a product of the nuclear reaction of α's on 94^{239} and is probably one of the following: 95^{242} (by *α,p* reaction) or 96^{242} (by an *α,n* reaction) or 96^{241} (by *α,2n* reaction). Other isotopes are possible, but these seem the most likely. The isotope 95^{241} from *α;n,p* is fairly probable.
>
> It is difficult to estimate the range of the α-particles from this new isotope very accurately, but it would seem to be about 4.65 ± 0.15 cm of air.

Incidentally, the range of 4.65 centimeters corresponds very well with the 6.1-MeV energy of the alpha particles from 96-242 known today.

Sample 49αA – #9, a mixture of a few hundred disintegrations per minute of the new long-range alpha activity and 12,000 disintegrations per minute of plutonium-239 alpha activity, was then

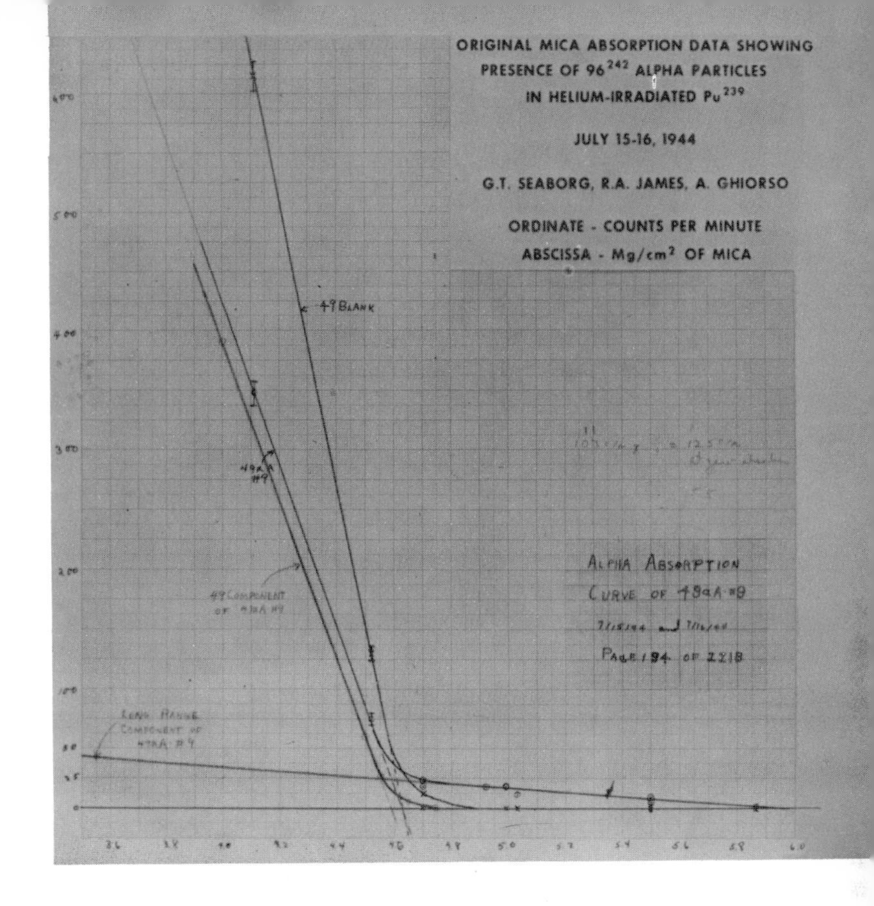

subjected to several additional oxidation cycles to further remove the plutonium-239. These were successful, and by Aug. 10, 1944, the sample was essentially free of plutonium-239. The sample was used to study the tracer chemical properties of the new isotope. The following is a verbatim extract from the Metallurgical Laboratory progress report (CS-2135) covering the period of August 1944:

> About 10 mg of Pu^{239} were bombarded to the extent of 40 μahr with 32 MeV helium ions in the Berkeley sixty-inch cyclotron. There was found in this material a new radioactivity, emitting alpha-particles of range about 4.7 cm, which seems to be due to an isotope of element 96 or 95. This new radioactivity is carried quantitatively by lanthanum fluoride, even in the presence of oxidizing agents such as dichromate or silver persulfate, indicating that this isotope cannot be oxidized to a +6 oxidation state in aqueous solution. The activity is carried by lanthanum oxalate in the presence of excess alkaline oxalate and not carried from acid solution by zirconium phenylarsonate, lead sulfate, or bismuth phosphate.

Time out to get married. . .

Let me interrupt the narrative momentarily to note that as I was studying the notebooks that recorded these technical events I came across the following entry, dated June 15, 1944, in Ralph James's notebook: "Time out to get married!" Apparently even discovering new elements was not allowed to interfere with love, but Ralph was back at work on June 19.

In the following months, it became increasingly apparent that this activity must be due to 96-242. The definitive evidence for this isotopic assignment came later through the observation of this same alpha activity as a product of the neutron bombardment of plutonium-239.

As the volume of the work and the complexity of the problem increased, we felt the need for more help. Early in September I asked Tom Morgan, a young chemist who had come to us from the University of Texas and who, like James, had distinguished himself in the investigations of the radiochemistry of plutonium, to join James, Ghiorso, and me in our search. He immediately joined Ralph in performing the chemical separations on a large (200 milligram) sample of plutonium which in August had received a very intensive deuteron bombardment in the St. Louis cyclotron. (Plutonium was now available in such quantities as the result of its production in the Clinton reactor.) The chemical procedure consisted in isolating a fraction presumed to contain element 95 by separating it from plutonium through repeated oxidation of plutonium to its soluble fluoride form and carrying the nonoxidizable element 95 as its insoluble fluoride on lanthanum fluoride.

There were some indications of an alpha particle of range longer than that of plutonium-239 in this element-95 fraction. On the basis of Ghiorso's absorption measurements, this alpha particle seemed to have a range of about 4.0 centimeters in air (compared to 3.7

It became increasingly apparent that this activity must be due to 96-242. . .

centimeters for plutonium-239), corresponding to an energy of 5.5 MeV. Tracer experiments were carried out during September and October using a total of about 100 disintegrations per minute of this 4.0-centimeter alpha emitter. These showed that it could be chemically separated from all the natural radioactive elements (lead and above) except actinium, thorium, and possibly protactinium. The range of the alpha particles seemed to be inconsistent with their being due to isotopes of any of these elements.

In retrospect, it seems we were catching a glimpse of 95-241 (whose alpha particles have a range of 4.0 centimeters) produced by a (d,n) reaction on the tiny amount of plutonium-240 (approximately $10^{-2}\%$) present in the plutonium. The observed intensity of 4.0-centimeter alpha particles (about 100 disintegrations per minute) was consistent with the now known yield for this reaction. Another source can now also account for the observed 4.0-centimeter alpha particles, suggesting that both sources made appreciable contributions to the observed intensity. Originally present in the plutonium (before the deuteron bombardment) was a very, very small concentration of plutonium-241 (approximately $10^{-5}\%$). As we now know, the plutonium-241 is long-lived, and thus its beta-particle decay would continually produce the daughter isotope 95-241 before, during, and after the deuteron bombardment. The amount of plutonium-241 and the time of its decay are consistent with the observed intensity of 4.0-centimeter alpha particles.

The presence of an alpha-particle emitter other than plutonium-239 could be clearly distinguished by alpha-particle-absorption experiments with mica absorbers. The following statement is quoted from our notebook (No. 727B) entry for Oct. 17, 1944:

> Among the heavy isotopes (Z = 80-94), Ac, Th, and Pa might follow the observed chemistry. This activity would have to be a new isotope of these elements, however, since none of the known ones would have this range or behave in this way (growth, decay). Element 95 would also be expected to have this chem[istry].

This activity had somewhat of a will-o'-the-wisp character, however, and the tiny supply was finally frittered away in the course of the chemical manipulations.

It was the neutron irradiation of plutonium-239 to a relatively large total exposure, first at Clinton Laboratories and then at the Hanford Engineer Works in Washington State, which led to the definite observation of an isotope of element 95 and the definite identification of the described product of the helium-ion bombardment of plutonium-239 as an isotope of element 96.

Two samples of plutonium-239, one of 4.4 milligrams and the

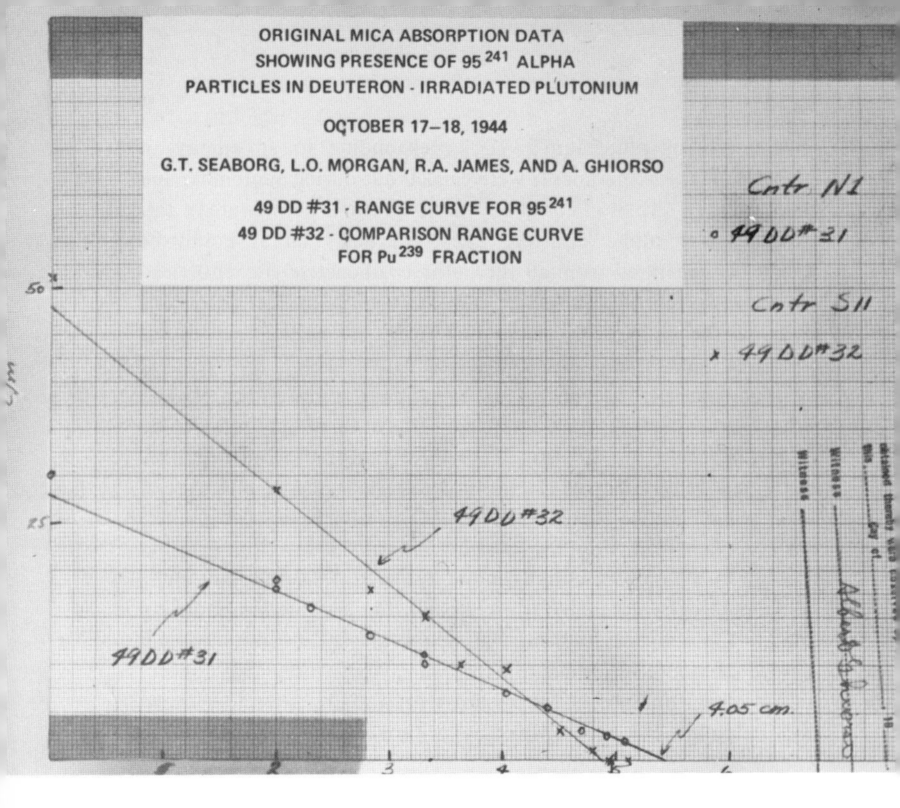

other 8.2 milligrams, were placed in the reactor at Clinton Laboratories on June 5, 1944. The first of these was removed on November 10 and returned to Chicago on November 19. This sample was put through the same type of chemical procedure described previously, in which the plutonium was oxidized to the VI oxidation state, whose fluoride is soluble, and insoluble lanthanum fluoride was precipitated with the intention of carrying any nonoxidizable isotopes of element 95 and 96 that might be present. After several such cycles had been completed, alpha-particle-absorption measurements were made on the fraction containing any nonoxidizable components to determine the ranges of the emitted alpha particles. A plot of these data on Dec. 6, 1944, showed that, in addition to the alpha particles from the plutonium-239 remaining in the sample, a component with alpha particles of longer range than those of plutonium-239 was present at an intensity of 200 disintegrations per minute.

The neutron irradiation of plutonium-239 to a relatively large exposure led to the definite observation of element 95 and element 96. . .

This was very exciting. As in the case of the long-range alpha particles observed the previous July in the nonoxidizable fraction isolated from plutonium-239 bombarded with helium ions and during September and October in plutonium bombarded with deuterons, here again was evidence that radiation from an isotope (or isotopes) of element 95 or 96 had been observed. This was confirmed during the following week when additional chemical cycles of the same type continued to remove the plutonium-239 while the intensity of the long-range alpha particles remained constant.

By the end of December, very careful continuing absorption measurements on the long-range alpha particles indicated the possible presence of two components.

The 8.2-milligram plutonium-239 sample was removed from the Clinton reactor on December 3 and returned to Chicago. This was also subjected to the now well-developed oxidation and lanthanum fluoride precipitation procedure to remove the plutonium-239 and concentrate the long-range alpha-particle emitter (or emitters). Again long-range alpha particles were found, and in greater quantity, as would be expected on the basis of the longer neutron irradiation.

Now we were sure that we had observed transplutonium isotopes, but there remained the task of putting together all the pieces of the puzzle. These began to fit together at about the turn of the year, early in January 1945. It was becoming increasingly clear that the long-range alpha particles from the neutron-bombarded plutonium-239 consisted of two components. But it was exceedingly difficult to define the two ranges with any accuracy.

Gradually we were able to conclude that the ranges of these two alpha particles in air were about 4.0 and 4.7 centimeters (compared to 3.7 centimeters for plutonium-239). It became clear to us, to our delight, that the 4.7-centimeter alpha particle was the same as that found as a product of helium-ion bombardment of plutonium-239. Thus it appeared that the 4.0-centimeter alpha emitter was due to 95-241 and the 4.7-centimeter alpha emitter was due to 96-242,

Now we were sure we had observed transplutonium isotopes. . .

---

produced in neutron-irradiated plutonium-239 by the following reactions:

$$^{239}\text{Pu}(n,\gamma)^{240}\text{Pu}(n,\gamma)^{241}\text{Pu}$$
$$^{241}\text{Pu} \xrightarrow{\beta^-} 95\text{-}241$$
$$95\text{-}241(n,\gamma)95\text{-}242$$
$$95\text{-}242 \xrightarrow{\beta^-} 96\text{-}242$$

The same isotope (96-242) was produced in helium-ion-bombarded plutonium-239 by the reaction:

$$^{239}\text{Pu}(\alpha,n)96\text{-}242$$

Aiding us in coming to this conclusion was the information we obtained from a 25-milligram sample of plutonium-239 that had received a relatively short neutron irradiation in November in the much higher neutron flux of one of the large plutonium production reactors at the Hanford Engineer Works. This sample, which was received on Jan. 15, 1945, was subjected to our by now standard chemical separation procedure. The element-95 and -96 fraction showed the same two-component long-range alpha particles that had been observed in the plutonium-239 samples irradiated in the Clinton reactor. Consistent with the higher neutron flux, the yields were much larger than had been obtained in the Clinton irradiations. In fact, it was an enormous quantity by our standards, of the order of 50,000 alpha disintegrations per minute.

A good summary of our understanding at that time can be seen in the following verbatim quotation from the Metallurgical Laboratory progress report (CS-2741) covering the period of February 1945:

> Some very interesting new alpha-radioactivity has been found, both in plutonium irradiated with neutrons in the Clinton pile and plutonium irradiated with neutrons in the Hanford pile. This alpha-activity exhibits just the sort of chemical behavior which has been predicted for the transplutonium elements. For example, it is carried by rare earth fluorides and it has not yet been possible to oxidize it to a state or states where its fluoride is soluble. It seems to be chemically separable from all the rest of

the 94 elements, and the best present interpretation is that it is due to elements 95 and/or 96. The alpha-activity is composed of two components, one of range 4.0 cm and the other of range 4.7 cm. A very attractive possibility is that the 4.0 cm alpha-activity corresponds to 95^{241} formed from the beta-decay of 94^{241} which comes from the reaction 94^{240} (n,γ) 94^{241}, and the 4.7 cm alpha-activity corresponds to 96^{242} (cf. reports CS-2124 and CS-2135) from beta-decaying 95^{242} following the reaction 95^{241} (n,γ) 95^{242}. The ratio of the yields in the Hanford as compared to the Clinton bombardment seems to be proportional to the second power of the total neutron irradiation, as would be expected on the basis of these particular isotopic assignments, although the accuracy of the estimation of the neutron fluxes is not sufficient to make this at all certain.

Soon after our results had been communicated to scientists at Los Alamos, they made a mass spectrographic identification in purified plutonium of a relatively volatile isotope with the mass number 241, which they identified with our 95-241. Since this could be extracted by volatilization from pure plutonium (from which the element 95 originally present had been removed in the purification process), we knew that the parent isotope, plutonium-241, must have a relatively long half-life (of the order of months or longer).

SUMMARY ALPHA PARTICLE ABSORPTION CURVES PRESENTED AT METALLURGICAL LABORATORY MEETING, CHICAGO, JANUARY 31, 1945

G.T. SEABORG, R.A. JAMES, L.O. MORGAN, AND A. GHIORSO

I - RANGE CURVE FOR 95^{241} AND 96^{242}
II - COMPARISON RANGE CURVE FOR Pu^{239}

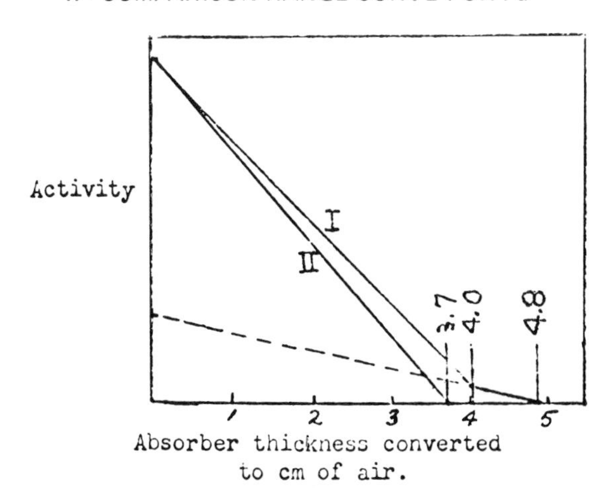

Our own work had not given us any information concerning the half-life of plutonium-241. To test for the presence of plutonium-241, we subjected 250 milligrams of purified plutonium produced in a Hanford reactor to our chemical procedure for the isolation of element 95. By March 20 we had a purified fraction containing the extracted 95-241. The intensity of the 4.0-centimeter alpha particles corresponded to about 30,000 disintegrations per minute. This established beyond doubt the identity of our 4.0-centimeter alpha emitter. Also, we now had a large sample with which to study the chemical properties of element 95, and, beyond that, we knew that we had an inexhaustible supply in our large stock of plutonium. Subsequent "milkings" led to continued extractions of the 4.0-centimeter alpha emitter from plutonium samples and we were able to establish the half-life of plutonium-241 (now known to be 13 years).

In March we had another bombardment of plutonium-239 with helium ions in the Berkeley 60-inch cyclotron. Again we chemically separated an element-96 fraction using our well-established procedures. Imagine, then, our surprise and consternation when the alpha particles emitted in this fraction had a range of 5.0 centimeters rather than 4.7 centimeters of the expected 96-242. However, we were soon able to solve this additional puzzle. It happened that the Berkeley cyclotron had been modified since our July bombardment, so that it now delivered 40-MeV helium ions rather than the previous 32-MeV energy. This meant that isotopes such as 96-241 and 96-240 would now be produced in predominating yields, due to $(\alpha,2n)$ and $(\alpha,3n)$ reactions, in addition to the previously produced 96-242. It was possible later to assign the 5.0-centimeter alpha emitter to the new isotope 96-240 and to detect the lower intensity alpha particles due to the 96-242.

A few months later, in July, the isotope plutonium-238 was identified as the alpha-particle-decay product of 96-242, thus lending further confirmation to this isotopic assignment. Decay measurements showed that 96-242 had a half-life of about five months (and 96-240 a half-life of about one month), but 95-241 exhibited no measurable decay (it is now known to have a half-life of 434 years). As soon as a sufficient quantity of 95-241 was available as a decay product of plutonium-241, it was irradiated with reactor neutrons to confirm the production of 96-242 by the reactions 95-241(n,γ)95-242 $\xrightarrow{\beta^-}$ 96-242. Soon thereafter W. M. Manning and L. B. Asprey measured the half-life of 95-242 as about 16 hours.

Of special interest is the fact that we were able to produce and

We were able to produce and identify the isotope 95-241 also by charged-particle bombardments. . .

---

definitely identify the isotope 95-241 also by charged-particle bombardments. Using helium ions from the Berkeley cyclotron to bombard uranium, we were able to identify plutonium-241 and its beta-decay daughter 95-241 produced by the reactions $^{238}U(\alpha,n)$ $^{241}Pu \xrightarrow{\beta^-} 95\text{-}241$. The significance of this lies in the fact that we were able to obtain declassification of this method of production much before the original method for the production of 95-241 (via the neutron irradiation of plutonium-239) could be declassified. Therefore my first announcement at a scientific meeting of the discovery of elements 95 and 96, in a paper presented at an American Chemical Society symposium at Northwestern University on Friday, Nov. 16, 1945 (published in *Chemical and Engineering News,* Vol. 23, page 2190, Dec. 10, 1945), described the production of element 95 by helium ions rather than by neutron bombardment.

As it actually turned out, the discovery of elements 95 and 96 was announced to the world for the first time prior to the symposium. This occurred on the "Quiz Kids" radio program on Sunday, Nov. 11 (Armistice Day), 1945. I happened to be a guest on this radio program, and one of the kids asked me if any new chemical elements had been discovered at the Metallurgical Laboratory during the war. Since the information had already been declassified for the symposium to be held the following Friday, I replied in the affirmative. The following is a verbatim transcription of this question and answer:

> Richard: Oh, another thing—have there been any other new elements discovered like plutonium and neptunium?
>
> Seaborg: Oh yes, Dick. Recently there have been two new elements discovered—elements with atomic numbers 95 and 96—out at the Metallurgical Laboratory here in Chicago. So now you'll have to tell your teachers to change the 92 elements in your schoolbook to 96 elements.

Soon after this, on Dec. 15, 1945, I made a guest appearance on the Watson Davis program, "Adventures in Science," which was a national network radio program at that time. The following is a

PERIODIC TABLE SHOWING ACTINIDE ELEMENTS AS MEMBERS OF A SERIES

Arrangement by Glenn T. Seaborg

1945

1 H 1.008																1 H 1.008	2 He 4.003
3 Li 6.940	4 Be 9.02											5 B 10.82	6 C 12.010	7 N 14.008	8 O 16.000	9 F 19.00	10 Ne 20.183
11 Na 22.997	12 Mg 24.32	13 Al 26.97										13 Al 26.97	14 Si 28.06	15 P 30.98	16 S 32.06	17 Cl 35.457	18 A 39.944
19 K 39.096	20 Ca 40.08	21 Sc 45.10	22 Ti 47.90	23 V 50.95	24 Cr 52.01	25 Mn 54.93	26 Fe 55.85	27 Co 58.94	28 Ni 58.69	29 Cu 63.57	30 Zn 65.38	31 Ga 69.72	32 Ge 72.60	33 As 74.91	34 Se 78.96	35 Br 79.916	36 Kr 83.7
37 Rb 85.48	38 Sr 87.63	39 Y 88.92	40 Zr 91.22	41 Cb 92.91	42 Mo 95.95	43	44 Ru 101.7	45 Rh 102.91	46 Pd 106.7	47 Ag 107.880	48 Cd 112.41	49 In 114.76	50 Sn 118.70	51 Sb 121.76	52 Te 127.61	53 I 126.92	54 Xe 131.3
55 Cs 132.91	56 Ba 137.36	57 La 138.92 58-71 See La series	72 Hf 178.6	73 Ta 180.88	74 W 183.92	75 Re 186.31	76 Os 190.2	77 Ir 193.1	78 Pt 195.23	79 Au 197.2	80 Hg 200.61	81 Tl 204.39	82 Pb 207.21	83 Bi 209.00	84 Po	85	86 Rn 222
87	88 Ra	89 Ac See Ac series	90 Th	91 Pa	92 U	93 Np	94 Pu	95	96								

LANTHANIDE SERIES	57 La 138.92	58 Ce 140.13	59 Pr 140.92	60 Nd 144.27	61	62 Sm 150.43	63 Eu 152.0	64 Gd 156.9	65 Tb 159.2	66 Dy 162.46	67 Ho 163.5	68 Er 167.2	69 Tm 169.4	70 Yb 173.04	71 Lu 174.99
ACTINIDE SERIES	89 Ac	90 Th 232.12	91 Pa 231	92 U 238.07	93 Np 237	94 Pu	95	96							

Naming one of the fundamental substances of the universe is done only after careful thought. . .

---

verbatim transcript of a portion of that radio program:

> Announcer: By the way, I'd like to know whether you have named these two new elements that you have discovered.
>
> Seaborg: Well, naming one of the fundamental substances of the universe is, of course, something that should be done only after careful thought. We have been faced with considerable difficulty in these cases because we have run out of planets. Naming neptunium after the planet Neptune, and plutonium after the planet Pluto, was rather logical. But so far the astronomers haven't discovered any planets beyond Pluto. So we'll have to go to some other method of naming.
>
> Announcer: What's that, Dr. Seaborg?
>
> Seaborg: This hasn't been decided yet. One possibility might be to rely on some property of these elements. We do have an idea for the naming of element 95 along these lines and may have a suggestion to offer pretty soon. And, by the way, you may be interested to know that we have received lots of suggestions. Some good and some not so good.
>
> Announcer: Well, Dr. Seaborg, perhaps some of the listeners to "Adventures in Science" will want to make suggestions as to the naming of the new elements. Will you be willing to have them write in their suggestions?
>
> Seaborg: Well, I don't promise to follow the suggestions. But it might be interesting to know what the public thinks about naming new chemical elements.
>
> Announcer: Very well, Dr. Seaborg. If you want to suggest names for new elements 95 and 96, just drop a postcard to Watson Davis, Science Service, Washington, 6, D. C. And to all those who write in, Mr. Davis will send them a free copy of the current issue of *Chemistry* Magazine which contains Dr. Seaborg's full technical paper and a new arrangement of the chemical Periodic Table.

We gave a great deal of thought to the naming of elements 95 and 96. My theory that they should be chemically similar to the rare-earth elements was being borne out to such an extent that we were finding it almost impossible to chemically separate them from these elements. Although we eventually succeeded in separating them, during the period of our futile efforts to do so, Tom continually referred to elements 95 and 96 as "pandemonium" and "delirium."

Names were finally suggested for elements 95 and 96 in the course of a talk I gave at the annual spring meeting of the American

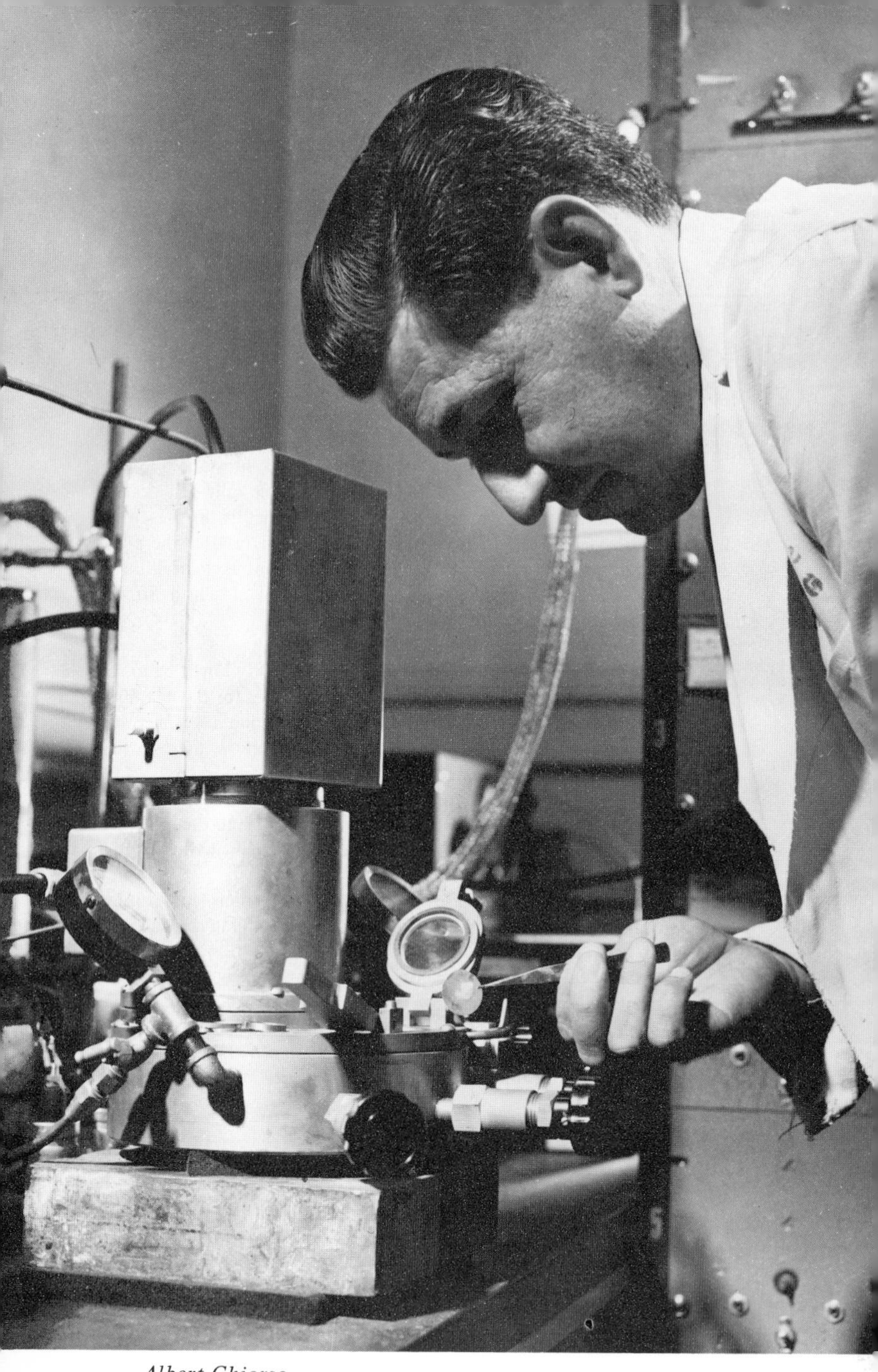

Albert Ghiorso.

Tom continually referred to elements 95 and 96 as "pandemonium" and "delirium"...

Chemical Society in Atlantic City on Apr. 10, 1946 (published in *Chemical and Engineering News,* Vol. 24, page 1192, May 10, 1946). Element 95 was given the name "americium" (symbol Am) after the Americas, in analogy to the naming of its rare-earth homologue europium after Europe. For element 96 we suggested the name "curium" (symbol Cm) after Pierre and Marie Curie, in analogy to the naming of its homologue gadolinium after Johan Gadolin.

This, then, is the story of the discovery of the elements with the atomic numbers 95 and 96. I hope that I have succeeded in recapturing for you some of the excitement, frustrations, and satisfactions that we experienced in the course of this scientific adventure. ■

# of ARTHUR H. COMPTO

Arthur H. Compton (right), shown here with Albert A. Miche son, at the time Compton was Dean of the Division of Physica Sciences, University of Chicago, a position he held when th first reactor was being built.

. . . Some Remarks and Reminiscences

At the Dedication of The Arthur Holly Compton Laboratory of Physics, Washington University, St. Louis, Missouri, May 4, 1966

■ It is a great honor and a real pleasure for me to speak to the distinguished group of administrators, educators, and scientists gathered here today to honor the memory of Arthur Holly Compton. No more fitting memorial to my friend and yours could be created than to have his name linked to a building like the one we are dedicating today. In this building, men of science—those already of great accomplishment and those aspiring to such—will share with Dr. Compton, in accordance with the basic purpose of Washington University, the building of an enduring civilization where men and women rise to the best that is in them.

As was well recognized by those who wished to honor the memory of Arthur Compton, it is particularly appropriate that this memorial should take the form of a laboratory of physics. Arthur was preeminently an experimental physicist. He had a burning desire to learn the facts of nature and to find their meaning. He was equally at home in the laboratory and in the development of bold theoretical concepts to interpret the experimental results obtained there. His unbounded confidence in both his experiments and his interpretation of the results brought him into sharp controversy with some of his contemporary fellow physicists on two occasions in two widely separated fields, X rays and cosmic rays.

Arthur Compton's experimental work on scattering of X rays, carried out here at Washington University in the early 1920's, laid the basis for the now famous Compton effect, the discovery of which earned him the Nobel Prize for Physics in 1927. But it

A burning desire to learn the facts of nature and their meaning. . .

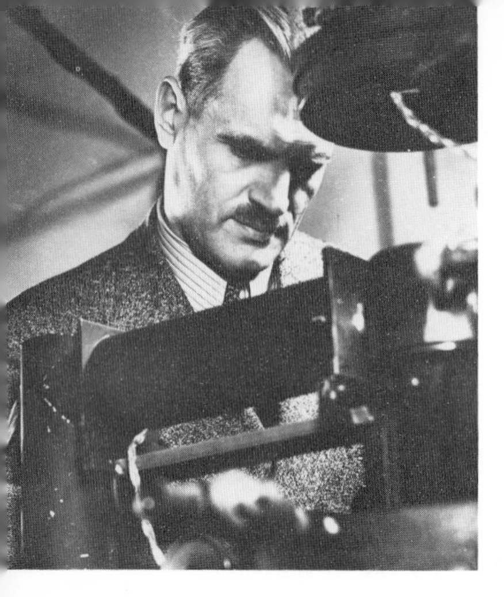

Equally at home in the laboratory. . .

and in the development of bold theoretical concepts. . .

required much experimental and theoretical checking before his concept of "billiard-ball" collisions between photons and electrons was accepted by some of his more skeptical colleagues.

At the height of his triumphs in the field of X rays, he decided that most of the fundamental work in this field had been accomplished and abruptly switched his attention to the new and more exciting field of cosmic rays. As might be expected, his first interest focused on trying to find out something about the nature of these rays. The principal question in this regard in the early 1930's was whether the cosmic rays consisted of uncharged photons (the "birth cry" of matter in far space) or of charged particles. It seemed to Arthur Compton that the best way to tackle the question would be to find out if the earth's magnetic field had any effect on the intensity of the cosmic rays. That was a big undertaking because the entire globe had to be the laboratory. But Arthur was equal to the occasion and organized systematic measurements of cosmic rays by widely spread expeditions on several continents. When his analyses of the global results convinced him that there was indeed a latitude effect of cosmic rays, showing that they consisted of charged particles rather than photons, he again found himself in controversy with those whose experimental results had revealed no latitude effect. And again his experimental results and conclusions were vindicated by subsequent experiments of others.

No doubt most of you are quite familiar with the story of Arthur Compton's life and the details of his major accomplishments, but perhaps on this occasion you will let me add a few footnotes to the Arthur Compton biography—footnotes derived from some personal reminiscences of my associations with Compton the scientist and Compton the man.

I count it among the most fortunate circumstances of my life that I had such a fine association with Arthur Holly Compton over so many years. I was familiar with Arthur's work in my college days, many years before I had the pleasure of meeting him and working with him.

My first meetings with Arthur Compton were casual ones. The first meeting that I would count as really important occurred in February 1942, and this one I recall vividly. At this time the Metallurgical Laboratory was just being set up at the University of Chicago. Plans were being made for the production of plutonium in quantity. A conference had been arranged to discuss the problems of the chemical extraction of plutonium after its possible production by the nuclear chain reaction operating with uranium. Arthur said he would like me to come from Berkeley to attend the conference, and I, of course, was pleased to have the opportunity.

I have vivid memories of this particular conference, for it was here that I was faced for the first time with the enormity of the problems connected with the extraction of plutonium from the large quantities of fission products that would be present in the uranium as a result of its undergoing the chain reaction. I remember that we discussed a number of possible methods of chemical extraction. In answer to a typically direct question from Arthur as to whether I thought the development of such an extraction procedure would be possible within the severely restricted time limitations involved, I replied that I thought it would. I also remember that I had some private doubts about such a fantastic proposition, but I did not

---

At the Metallurgical Laboratory plans were being made for the production of plutonium. . .

think it appropriate to express any lack of confidence in my part of the job in the face of Arthur's tremendous enthusiasm about the whole project.

I believe that Arthur Compton may have had far more faith in me at that time than perhaps I had in myself. Reading his autobiography, *Atomic Quest,* some years later, I came across the following passage relating to the plutonium extraction question:

> I went with Bush and Conant to lunch at the old Cosmos Club. In the table conversation I remarked that we should give further thought to the production of plutonium as an alternative to the separation of uranium 235. In spite of the unknown difficulties in establishing a controlled nuclear chain reaction, did not the advantage of chemical extraction in the case of plutonium instead of isotopic separation in the case of U-235 make this process a worthy competitor? Bush called attention to the obvious uncertainties that lay in putting into production a kind of process that was completely unknown to industry and which had yet to be shown possible in the laboratory. Conant added that, even if we could produce the plutonium, we knew almost nothing of its chemistry. Even when we had this knowledge the task of extracting the plutonium from the uranium would be greatly complicated by the intense radioactivity. It would take years to get the chemical extraction process in operation. This was Conant, the expert chemist, speaking from experience.
>
> "Seaborg tells me that within six months from the time the plutonium is formed he can have it available for use in the bomb," was my comment.
>
> "Glenn Seaborg is a very competent young chemist, but he isn't that good," said Conant.
>
> Actually the time from pulling the last activated slugs from the Hanford piles until the metallic plutonium that they contained was ready for use at Los Alamos for the first bomb turned out to be hardly two months.

That is the end of the quote from Arthur's book, and I am still grateful to him for his confidence.

It was only about two months after my conference with him that Arthur asked me to move from Berkeley to join him at the Metallurgical Laboratory to take charge of the work that would be required to develop such a chemical extraction process for plutonium. I remember that I arrived in Chicago on Apr. 19, 1942, which happened to be my thirtieth birthday. It also happened to be an unusually dark and dreary day, and I recall having some misgivings as to my judgment concerning the change in climate that I had elected.

My relations with Arthur during the next three years, while the many difficult problems in the production and extraction of plutonium were being worked out, were very close. He was the type of person whose door was always open to discuss the many problems involved.

I could tell you many, many stories about the Metallurgical Laboratory days, but I do not feel that I should spend my time here

The time from pulling the last activated slugs from the Hanford piles until the metallic plutonium they contained was ready for Los Alamos turned out to be hardly two months. . .

Chemical separation building under construction at Hanford. The completed building was 800 feet long, 65 feet wide, and 80 feet high. Here the process developed by Stanley G. Thompson and Metallurgical Laboratory chemists would be used to separate plutonium from irradiated uranium.

University of California at Los Angeles chemists assembled at the Metallurgical Laboratory on July 25, 1944: (left to right) Leonard I. Katzin, Zene V. Jasaitis, Nathalie S. Baumbach, Harlan L. Baumbach, Glenn T. Seaborg, Stanley G. Thompson, J. Leonard Dreher, and Fred W. Albaugh.

today in such an occupation. One incident, however, is worth relating at this point because it brings to mind Arthur's tremendous powers of concentration. On a cold, stormy, wintry night—I believe it was in January or February— he walked over from his office in Eckart Hall on the University of Chicago campus to my office in the New Chemistry building on Ingleside Avenue adjoining the campus. After our talk in my office on a subject I cannot now recall, we continued our conference as we walked down Fifty-seventh Street on the way to our homes. The going was treacherous, and I recall that at one point he slipped on the ice and fell head over heels to the sidewalk. He got up with amazing speed and, totally unconcerned, continued without noticeable interruption to the end of the sentence he had had in progress before the incident.

I remember that during that period at the Met Lab my wife, Helen, had an especially close relation with Betty Compton in connection with Betty's many activities. Betty was Treasurer of the Chicago Metropolitan Y.W.C.A., an organization with a sizeable budget, and she interested a large number of "the Project" wives in her project.

The relation of Helen and myself with Arthur and Betty Compton did not cease with the successful completion of the tasks of the Plutonium Project. We have had many contacts since that time.

After Arthur took up his duties as Chancellor here at Washington University, I visited the campus on numerous occasions. Immediately after the war he brought Joseph Kennedy, my close associate at Berkeley who had participated with me in the discovery of plutonium, to the University as the chairman of the Chemistry Department. Arthur Wahl came to the University as a member of Kennedy's department at about the same time and is still here, his fine reputation as a nuclear chemist having been enhanced recently by receipt of the American Chemical Society's Award for Nuclear Applications in Chemistry. As you know, Wahl worked as a graduate student with Kennedy and me in the discovery of plutonium, and surely Art Wahl's is one of the historic Ph.D. theses of all time. So, because of my various associations here, I have had a number of reasons to visit Washington University in the past and have taken advantage of my opportunities to do so.

I might add that there is also another important connection between Washington University and plutonium. During the early days of the war and of the Plutonium Project, the Washington University cyclotron was used for the preparation of important amounts of plutonium. These were weighable amounts, hundreds of

Leaders in atomic energy meet at a 1946 dinner to honor Arthur Compton's inauguration as Chancellor of Washington University in St. Louis: (seated, from left) Maj. Gen. Leslie Groves, Vannevar Bush, Enrico Fermi, Brig. Gen. Kenneth Nichols, George Pegram, and Lyman Briggs; (standing, from left) Charles Thomas, James Conant, Chancellor Compton, Eger Murphree, and Crawford Greenewalt.

micrograms, huge by the standards of those days and very important to the success of the Project.

In thinking back on my associations with Arthur Compton, I remember particularly that he and I served as members of the Historical Advisory Committee of the Atomic Energy Commission and that Arthur was an especially great aid to Hewlett and Anderson in the advice he gave them with respect to Volume I of the history of the Atomic Energy Commission *The New World.* I can recall vividly a meeting of the Historical Advisory Committee held in Richland, Washington, in the summer of 1960. At a huge public gathering held in the Village Theater during the evening, Arthur and I shared the platform to reminisce about the wartime Plutonium

At Chicago in October 1946, Arthur Compton receives a certificate from the scientists of the Metallurgical Laboratory: "---in appreciation of his broad vision, his courageous and inspiring leadership, and his unfailing sympathetic support in the research and development required for the production of plutonium---." Also shown here are Farrington Daniels and Col. Arthur V. Peterson.

Project and to share our thoughts on its present and future consequences. The discussion went on to explore the rich area of the philosophy of science, and it showed Arthur at his best.

In addition to these high moments, there were many other times of pleasant personal association and times when we were in contact on a variety of matters. In connection with my duties as Chancellor of the University of California at Berkeley, I arranged, with the help of Dr. Edward W. Strong, an outstanding professor of philosophy on the Berkeley campus, a visiting professorship for Arthur to discuss the relation between science and philosophy. These two eminent scholars were most interested in developing this field. Arthur and Ed struck up an immediate and firm friendship, and Arthur was actively engaged in this field of endeavor to the last day of his life.

Arthur was always interested in athletics. He played football in college—most if not every game was attended by Betty—and he always maintained his enthusiasm for the game. He and Betty attended a football game at Berkeley with us when I was chancellor, and, although I fancied myself somewhat of an expert, his greater understanding of the fine points of the game was readily apparent.

His leadership in the Plutonium Project has become a classic example of the team approach to difficult and complex problems. . .

Arthur loved children, and they shared his feeling. I remember the trouble he went to when our youngest child was a baby in taking color photographs of her. His prints are among the finest pictures we have.

Compton recognized that we live in an interdependent world where it is necessary for all of us to adapt ourselves to the rapid change brought about by science and technology. He recognized also that the scientific laboratory had outgrown the stage when good work could be done with simple collections of glassware, microscopes, agar plates, and the like. It was clear to him that accomplishments for the betterment of mankind were possible only by full cooperation among specialists in many fields, each working to enable the group to achieve a common goal. His leadership in the Plutonium Project has become a classic example of the team approach to difficult and complex problems. At the same time, however, he realized the importance of the personal contribution of the individual, either working alone or as a team member, in achieving a common goal. He saw in the success of the team approach to the solution of scientific problems a valuable pattern for society as a whole to follow and recognized the search for truth as a cooperative process in which every step on the road to the solution of a problem presupposes the validity of all previously taken steps and prepares for all new steps to be taken in the future. Perhaps his philosophy about the role of the individual and the interdependence of individuals in today's society is best summed up in a statement he wrote about their relation in terms of his own career:

> As I think back to these early experiments, one matter strikes me forcibly. It is the contrast between my first airplane experiments, which were done entirely with my own hands and with a minimum of contact with others working in the same field, and my last major engineering or scientific job, which was the atomic reactor, where thousands of people were engaged using the best available tools, and my part consisted largely in organizing the activities and the thoughts of the scientists who were concerned with getting the job done. In the thirty-three years that intervened my own scientific work, like that of the nation, went through

all the stages from primitive pioneering, where one's own individual skill and resourcefulness, not only in ideas but also in handcraft, was the basis of success, to what has now become a vast enterprise in which each individual contributes a small but expert part, but in which, nevertheless, the originator of the idea has still to take the responsibility for it.

The example which Arthur Compton set for us on the high value of the search for truth inspires us all. The need to satisfy the ever-increasing number of people wanting an education and Compton's personal dedication to the development of the inherent value of every person led him to take on the important position of Chancellor of Washington University. He recognized that the ideal of a great American university rests entirely on the theory of the dignity of the human spirit. This ideal is the very embodiment of democracy; it emphasizes the individual.

Today we see this ideal embodied in the Arthur Holly Compton Laboratory of Physics, an edifice that is not only the culmination of the efforts of its material constructors but also a monument reminding us of the theory of the dignity of the human spirit, of the concept of truth, and of the high value of the search for truth. Those

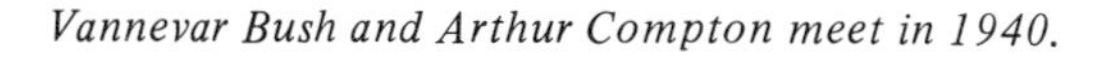

Vannevar Bush and Arthur Compton meet in 1940.

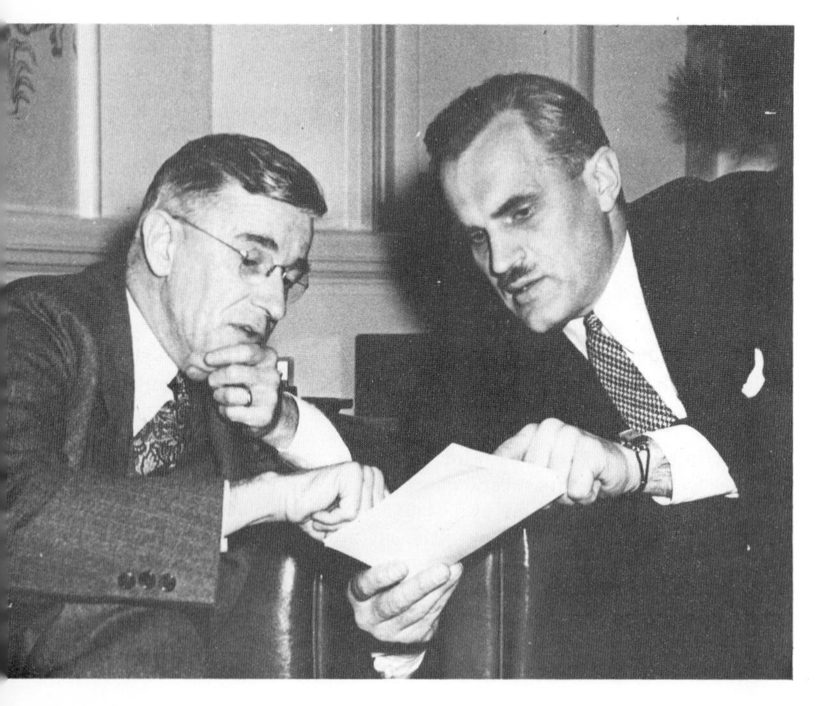

A 1940 meeting on the Berkeley campus of the University of California to discuss the proposed 184-inch cyclotron: (left to right) Ernest O. Lawrence, Arthur H. Compton, Vannevar Bush, James B. Conant, Karl T. Compton, and Alfred Loomis.

who work in its halls will realize, from the great example of Arthur Holly Compton, that their obligation is to accomplish those scientific tasks that are new and hard and deep, because such deeds are essential to the physical, mental, and moral advancement of men and women everywhere. ■

The Heritage of

ERNEST O. LAWRENC

. . . Dreamer in Action

Ernest O. Lawrence, inventor of the cyclotron, at the ion source of the 184-inch cyclotron in 1948.

The Lawrence Hall of Science, University of California, Berkeley.

At the Dedication of the Ernest O. Lawrence Hall of Science, University of California, Berkeley, California, May 20, 1968

■ It is an understatement to say that this is a day of great pleasure for me. It is always good to be in Berkeley, which I still regard as home. I am delighted to join in this ceremony with members of the Lawrence family and with the friends, associates, colleagues, and admirers of the late Professor Lawrence. Above all, I am honored beyond words to have been chosen to dedicate this Ernest O. Lawrence Hall of Science to which many people across the nation, in many walks of life, have given so generously of themselves and their substance. The breadth and depth of feeling embodied in this memorial testifies to the inspiration Ernest Lawrence projected to those who knew him and to others who, although unacquainted with him, understood his significance to our times.

As I prepared for today's dedication, my mind went back nearly a decade to the origins of this project. I find that the memory is still sharp of the shock and dismay that followed word of the premature passing of Professor Lawrence. Almost immediately thereafter the Regents of the University of California renamed the great Radiation

---

A seemingly inexhaustible supply of energy projected on a flame of faith that never seemed to flicker. . .

Independence, initiative, and a capacity to see quickly what was important in matters that interested him. . .

Laboratory after its founder, and this same body established a committee to consider a suitable additional memorial to Dr. Lawrence. Subsequently, the Atomic Energy Commission, with the approval of President Eisenhower, established the Ernest O. Lawrence Memorial Awards to reward high achievements in nuclear research by young scientists. This year's awards will be presented later in this program.

As Chancellor of the Berkeley campus, I had the privilege of being involved in the early efforts to implement the desire of the Regents to establish a memorial. Since going to Washington early in 1961, I have continued to follow the evolution of the project as a member of the Advisory Committee. The two dominant considerations in developing the concept of this Hall of Science have already been outlined by Dr. Harvey White. First, we sought to perpetuate in some dynamic and socially useful form the personality, the ideals, and the work of Ernest Lawrence. Second, we sought to do this in a way appropriate to the university's functions as a great teaching and research institution.

I should like to speak particularly to the first consideration, namely, Ernest Lawrence himself. A descendant of educated Norwegian immigrants, he was born on Aug. 8, 1901, in the small town of Canton, South Dakota, on the Big Sioux River. There, in the agricultural expanse of the Great Plains, he grew up. His character was shaped by traditional American values found in the home, the church, the school, and the land itself—all leavened by the liberal perspective of an educator father.

All through his life Ernest was special. As a child, he was wiser and older than his years—a youth ahead of his time. Early in his life he showed unusual qualities of independence, initiative, and a capacity to see quickly what was important in matters that interested him. From some unknown source he drew on a seemingly inexhaustible supply of energy, which he expended so lavishly in achieving his goals that he caused wonder among observers. His energy was projected on a flame of faith that seemed never to flicker—faith in himself, in others, in human ingenuity, and in the future.

Throughout his life his eyes were fixed on the horizon. He visualized large devices and projects, seemingly unattainable to other men, that would serve science, the nation, and mankind. He was a dreamer, but a dreamer in action. Once he had formed a concept, he wasted no time in mobilizing his diversified resources in massive attacks that brought obstacles crashing down, creating realities where without him only bemusing fantasy might have persisted. He showed time and again that the word "impossible" should not be used with abandon. Hardly pausing after each triumph, he rushed impatiently forward toward a new dream.

Lawrence's concept of the cyclotron stamped him as an intellectual genius. His transformation of the concept into giant machines of increasing power to explore the nucleus of the atom demonstrated genius of broader scope—and accounts for his wide appeal. In the course of these transformations, Dr. Lawrence, bold, adventurous, and unfettered by the tradition that an academic scientist used only the tools he could make himself, brought engineering and technology into the basic science laboratory. The utility of his accelerators and their products cut across a broad spectrum of scientific fields, and he joyously promoted these uses. Physicists, chemists, biologists, physicians, engineers, and agricultural scientists rubbed elbows and traded ideas in his laboratory. Here at Berkeley interdisciplinary research came into its own, nuclear medicine was born, and modern nuclear chemistry emerged. The pattern of large-scale basic research now seen in the national laboratories in this country and abroad was formed. The potentials for doing scientific work were enormously expanded. Although Dr. Lawrence's central interest was the expansion of knowledge, he believed deeply in making science useful to man; thus he was a prime mover in the early demonstration of the practical benefits of nuclear energy. Lawrence's widely emulated tools and ways of conducting scientific work accelerated progress and our present scientific revolution.

Beginning in the early 1930s and continuing today, a steady stream of discovery has flowed from his laboratory, a flow so massive as to change the landscape of science. Even a summary of these scientific achievements would be too voluminous to include here. However, I believe just a few examples will illustrate the impact of Professor Lawrence's laboratory on science. In choosing examples I decided to look at items falling at 10-year anniversary intervals beginning in 1938. Even so I must be selective.

Among other developments in 1938, the cyclotron, having already shown its power for exploring the nucleus, demonstrated in a broad way that nuclear science would yield extensive practical

Lawrence's concept of the cyclotron stamped him as an intellectual genius. . .

Original cyclotron equipment used by E. O. Lawrence and N. E. Edlefsen in 1930.

Stanley Livingston and Ernest Lawrence in front of the 27-inch cyclotron in the old Radiation Laboratory in 1934.

The University of California 60-inch cyclotron soon after completion in mid-1939. Key figures in the development: (left to right, bottom) Donald Cooksey, Dale Corson, Ernest Lawrence, Robert Thornton, John Backus, and Winfield Salisbury; (top) Luis W. Alvarez and Edwin M. McMillan.

His transformation of the concept into giant machines of increasing power demonstrated genius of broader scope. . .

The 184-inch cyclotron completed in 1946 with an energy for deuterons of 200 million electron volts.

Startup of the 184-inch cyclotron, November 1946: (left to right) Robert Thornton, Ernest Lawrence, and Edwin M. McMillan.

He visualized large devices and projects seemingly unattainable to other men. . .

The Bevatron, another Lawrence giant particle accelerator, began operation in 1954.

applications for the benefit of man. As we look back, the applications in medicine that year are particularly striking. The cyclotron had proved to be the best instrument for discovering radioisotopes, and it was also the only means in the 1930s for producing useful quantities of these substances. Ernest Lawrence himself had early produced sodium-24 in quantity, and he and his associates had demonstrated its utility in studying metabolism. In 1938 one of the notable isotopes discovered at Berkeley was iodine-131. Immediately the late Dr. Joseph G. Hamilton, a physician who had joined the staff, together with colleagues at the Medical Center in San Francisco, began to develop techniques for using radioiodine for the diagnosis and treatment of thyroid disease. Today, in many medical centers in this country and in other advanced countries, radioiodine has replaced other techniques for detecting thyroid diseases and for treating hyperthyroidism (over-activity of the thyroid gland). Half a million radioiodine cocktails are administered annually in this country alone. Iodine-131 today probably provides man with the most extensive direct benefit of nuclear energy in medicine. It is a great source of satisfaction to me to have participated, with J. J. Livingood, in the discovery of this isotope, using the historic 37-inch cyclotron that will be displayed in this Hall. Hundreds of other radioisotopes have been discovered at Berkeley. Another notable one identified here in 1938 is iron-59, which Dr. John H. Lawrence, Director of Donner Laboratory, and other scientists have used to expand greatly our knowledge of blood metabolism and disease. One isotope deserving special mention because it has the greatest value and the widest application to science, as well as to many areas of industry, is carbon-14, discovered in 1940 by the late Dr. Samuel Ruben and Dr. Martin Kamen, now of the San Diego campus of the university. Carbon-14, among other things, has revolutionized biology.

Another application of nuclear science to medicine originating in 1938 was the first use of an accelerator beam to treat cancer, initiated by Dr. John Lawrence, with colleagues from the San Francisco Medical Center. In recent years, with more powerful cyclotrons, Dr. Lawrence and his Donner colleagues have successfully used particle beams to treat acromegaly, Cushing's disease, and other disorders rising from malfunction of the pituitary gland. There is a continuity to this work that is important to the future. One of the major potential uses of the Omnitron, an accelerator concept that has recently emerged from the Lawrence Radiation Laboratory, will be to explore the treatment of cancer and other diseases by intense beams of particles heavier than those now available.

Another banner anniversary year was 1948, which saw one of Lawrence's many big dreams realized. When Ernest Lawrence conceived the 184-inch cyclotron before World War II, he hoped to create mesons, particles that one day may also be used for cancer treatment. At that time mesons could be found only in cosmic rays. They were believed to supply the "glue" holding the atomic nucleus together. If these particles could be produced in the laboratory under controlled conditions, perhaps it would be possible to understand the nuclear force. The war intervened, and during that time Dr. Lawrence received an assist toward his goal by the discovery of a new principle of particle acceleration by Dr. Edwin M. McMillan, present Director of the Lawrence Radiation Laboratory, and V. Veksler of the USSR. The principle extended the energy potential of accelerators, including the 184-inch cyclotron. The

Cesare Lattes and Eugene Gardner in early March 1948 setting up a key discovery experiment on the target probe of the 184-inch cyclotron. In this experiment several months previously they were the first to observe man-made mesons.

result was the detection of man-made mesons for the first time in 1948. This marked the beginning of modern high-energy physics in the laboratory, and since then Berkeley and other accelerators have revealed the existence of a host of previously unknown particles of matter. This event is the starting point for the present revolution in our concepts of matter.

When we come to 1958—April 1958, to be exact—we see another notable discovery that is representative of a whole body of knowledge created in Berkeley, the discovery and study of artificial elements. Prior to this date Berkeley scientists had filled in the Periodic Chart beyond uranium from element 93 through element 101, with collaboration from other laboratories in the case of elements 99 and 100. Incidentally, many of you know that the room in Gilman Hall on the campus where the nuclear energy fuel plutonium was discovered has been designated a National Historic Landmark. In April 1958 Albert Ghiorso and his colleagues in the laboratory discovered element 102. It is quite appropriate, incidentally, that Ghiorso and his associates, when they discovered element 103 in 1961, established their own memorial to Professor Lawrence by giving this element the name lawrencium.

Lawrence was awarded the Nobel Prize in 1939 and the Enrico Fermi Award in 1957.

Perhaps this brief and necessarily selective reflection on Professor

In September 1942 the leaders, then, of the atomic bomb project met at the Bohemian Grove in California as guests of Ernest Lawrence. At this meeting they decided to proceed with a production plant for the electromagnetic separation of uranium-235 which Lawrence had been developing in his Berkeley laboratory: (left to right) Maj. Thomas Crenshaw, J. R. Oppenheimer, Harold C. Urey, E. O. Lawrence, James B. Conant, Lyman J. Briggs, E. V. Murphree, A. H. Compton, R. L. Thornton, and Col. K. D. Nichols. Oppenheimer, Thornton, and Nichols were consultants to the Committee.

Lawrence and his impact on science will explain how the purposes of the Hall of Science, which have been stated by Dr. White, were derived. The Hall is centrally concerned, as was Lawrence, with the young, with science, and with the future. This living memorial is developing new and better ways of teaching science to the young and is placing these methods in the hands of the teachers of youth. The Hall seeks to bring a better exposure to science to students not only in California but all over the nation. The example of Professor Lawrence suggests the importance of this national perspective. A small town on the Great Plains seems at first glance to be an unlikely place from which one of the giants of modern science would emerge. Yet we never know where genius will be found, and we must try to encourage the flowering of genius in science everywhere.

To me this Hall almost exceeds in reality the dream we had nearly a decade ago of a living memorial that would do justice to the man for whom it is named. Its goals are important to all of us, and in physical aspect and surroundings it is inspiring. Above is the Space Sciences Laboratory, which beckons toward great future adventures in a vastness that is hard to comprehend. Immediately below is the Lawrence Radiation Laboratory, founded by the man we honor today, which continues as it has for nearly four decades its vigorous explorations of the unimaginably small—the atom and its components. At the bottom of the hill is the great university center embracing the whole spectrum of intellectual inquiry, and beyond is the magnificent San Francisco Bay, garlanded with cities and bridges. With continued devotion by those dedicated to the ideals embodied in this memorial, the Ernest O. Lawrence Hall of Science will make an enduring contribution worthy of the man whose memory it honors. ■

J. ROBERT OPPENHEIM

His Public Service and Human Contributions

At the J. Robert Oppenheimer Memorial Session, American Physical Society Meeting, Washington, D. C., Apr. 24, 1967

■ It is an honor and a privilege to have this opportunity to speak to the American Physical Society about Robert Oppenheimer, who was one of this society's most admired, respected, and distinguished members, as well as one of its most able presidents. I also feel somewhat honored to be asked to represent the chemists on this occasion—if I may indeed be classed as a chemist. But, in making my remarks in tribute to Oppie, I will try to uphold the honor of those chemists who did work with him. For those physicists who may not have known it—and to bolster my own confidence—I should recall that early in his life there was a time when Oppie almost chose chemistry as a career.

Much has been said and written about Robert Oppenheimer by the many friends and students with whom he shared his life—his knowledge and wisdom, the wealth of his personal warmth, and the wide range of his worldly interests. I hope I can avoid repeating too much of what others have said and can add a few new thoughts. In reviewing some of Oppenheimer's contributions, I would like to begin with my personal recollections of him. I was fortunate enough to have known and worked with him over a long period of time,

His magnetic —really electric— personality. . .

The early years of nuclear discovery and development when we shared some of the excitement . . .

particularly in the early years of nuclear discovery and development when we shared some of the excitement of those historic days.

I first met Robert Oppenheimer when I went to the University of California as a graduate student in chemistry in 1934. Oppie, who had then just passed his thirtieth birthday, was an associate professor of physics, dividing his time between Berkeley and the California Institute of Technology in Pasadena. I must confess that he made a terrific impression on me—an impact I never quite got over in the following thirty-odd years of acquaintance with him. And I have the feeling that his memories of a gangling, young, naive, would-be nuclear chemist may have continued to color his view of me long after I pictured myself as having reached a moderate stage of maturity.

I am afraid that I may have manufactured occasions that made it necessary for me to consult with him regarding my research problems. In retrospect, I do not see how my problems could have been of great intrinsic interest to him, but I cannot recall any occasion when he was at all unwilling to help. One particularly puzzling riddle, a real one in this case, concerned the results of the irradiation of various elements with fast neutrons in the million-electron-volt energy range, which I was doing in the mid-1930s with David C. Grahame, also a graduate student at that time. The Japanese physicist S. Kikuchi and his coworkers had observed in such experiments the production of electrons in the million-electron-volt energy range, and they attributed these to some unusual direct interaction of the fast neutrons with orbital electrons. Grahame and I preferred the view that these electrons were the internal conversion products of gamma rays produced in nuclei that were excited by the inelastic scattering of the neutrons, at that time an unobserved—or at least unproved—process. This interpretation presented a problem, however, because the experimental results suggested internal conversion coefficients much higher than had generally been observed up to that time. This is one of the riddles I presented to Oppie, and I believe that I succeeded in intriguing him. The explanation came some time later as a result of the recognition

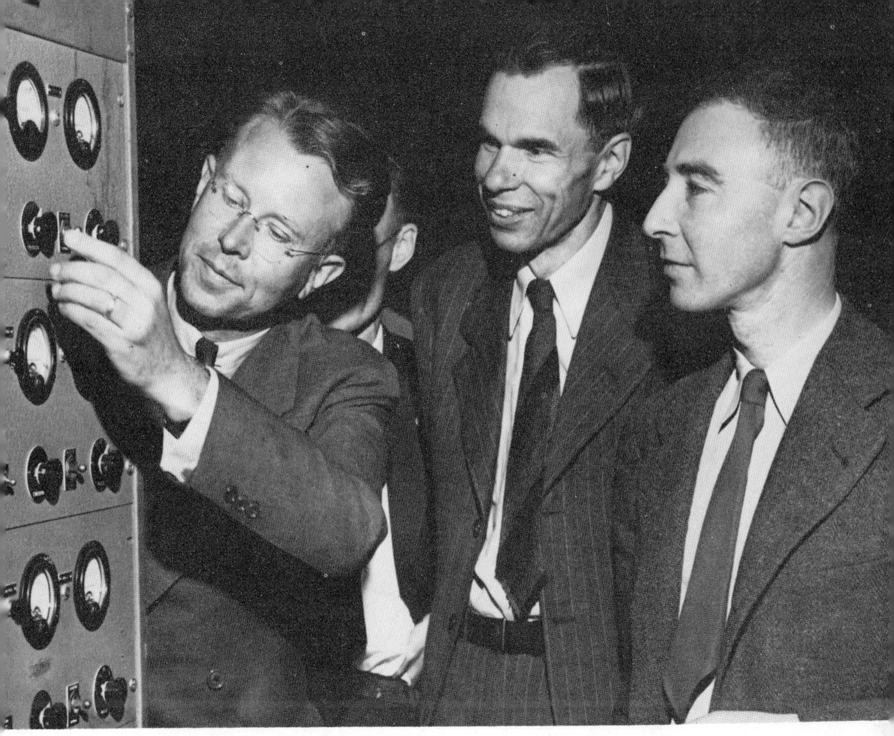

Lawrence, Seaborg, and Oppenheimer in early 1946 at the controls to the magnet of the 184-inch cyclotron, which was being converted from its wartime use to its original purpose as a cyclotron.

of the role of spin change in slowing down gamma-ray transitions and increasing their internal conversion.

I imagine I had one difficulty with Oppie which all who sought his advice had, that is, facing his tendency to answer your question even before you had fully stated it. I recall taking great pains in formulating my questions to him so that I could get the main thrust of my thoughts as early as possible into every sentence.

I particularly remember Oppie's role in Physics Department seminars. Everyone turned to him for explanation of their experiments in nuclear physics, and his electric personality certainly contributed to our fascination and satisfaction with his performance. I remember particularly that I was present at the seminar in January 1939 when the newly arrived results of Hahn and Strassmann on the splitting of uranium with neutrons were excitedly discussed. I do not recall ever seeing Oppie so stimulated and so full of ideas. As it turned out, I was thus privileged to witness his first encounter with the phenomenon that was to play such an important role in shaping the future course of his life.

The year 1939 changed many things. . .

Oppenheimer's home in Los Alamos.

Party at Dorothy McKibben's home. Isadore I. Rabi, Dorothy McKibben, Robert Oppenheimer, and Victor Weisskopf.

Robert Oppenheimer and Gen. Leslie R. Groves on the Trinity site.

After the war, the Army–Navy "E" Award was presented to the Los Alamos Laboratory.

The first Atomic Energy Commission at Oak Ridge, Nov. 12, 1946: (left to right) William W. Waymack, Lewis L: Strauss, David E. Lilienthal, Robert F. Bacher, and Sumner T. Pike.

Members of the General Advisory Committee visit Los Alamos. Here, having just landed at the Sante Fe airport, Apr. 3, 1947, are (left to right) James B. Conant, Robert Oppenheimer, Gen. James McCormack, Hartley Rowe, John H. Manley, Isadore I. Rabi, and Roger S. Warner. Manley was the Committee's secretary. McCormack and Warner were members of the Commission's staff.

As Viki Weisskopf said, the year 1939 changed many things, and Viki has given us an illuminating account of Oppenheimer's marvelous leadership at Los Alamos. Although I spent the war years at the Metallurgical Laboratory in Chicago, my contacts with Oppenheimer continued. He visited me on many of his trips to Chicago, and our discussions continued to impress me with his enduring interest in chemistry, an interest that was perhaps not so well known. We discussed not only the problems with the chemical purification of plutonium required for its successful use as an explosive but also the various chemical methods under investigation for the separation of plutonium from uranium and fission products at Hanford.

My closest association with him came when I served as a member of the Atomic Energy Commission's General Advisory Committee (GAC) during the first 3½ years of its existence. The GAC played a very important role in those formative years of the Commission, and

In no other era had the world seen such a transfer of theory into application. . .

Oppenheimer, as Chairman of the GAC, was the architect of that role. The GAC, whose other members at that time were Conant, DuBridge, Fermi, Worthington, Rabi, Rowe, and Smith, had the responsibility of setting the initial course of the AEC's military program and guiding the first ventures in the peaceful uses of nuclear energy.

I recall how impressed I was with Oppie's leadership of the committee. During the 3½ years we met in about 20 sessions. These three-day-long meetings were usually held over the weekends when we had the time, or took the time, to leave our other duties. At the conclusion of each session, when the AEC Commissioners came in to review our work, Oppie presented a brilliant and masterful summary of the proceedings. I know that my fellow members of the GAC remember with me that this was pure Oppenheimer at his very best. I regret that tape recordings were not made of these eloquent summations of our deliberations, for I believe that these would be better than the written record that followed and would provide fascinating historical material.

It is not generally appreciated how much of Oppenheimer's efforts in those early GAC meetings went toward strengthening the Commission's and our nation's position in national defense. He devoted great effort to programs that strengthened the position of the Los Alamos Laboratory, and he emphasized the priority of plutonium production at Hanford.

During those early GAC days, Oppie also showed his great desire to foster the peaceful role of the atom. Like most of us, he wanted to see the early development of nuclear power. But, also like most of us at that time, he was somewhat overly pessimistic about the possibilities of rapid growth in this area. Judged by the current activity in nuclear power, his early report on the outlook for developing civilian nuclear power did not anticipate the possibilities being realized today.

During his leadership of the GAC, Oppenheimer spearheaded the move for strong AEC support of fundamental research. In no other era of human history had the world seen such a transfer of theory into application as in the events of the Manhattan Project. Perhaps better than any other person, Oppenheimer, who had overseen so much of this project, saw this transfer take place. The realization of the future implications of fundamental research had a most profound effect on a man of his depth and philosophical insight. He saw the dawn of a new age of science and knew that the government's relation to science could never be the same again. Therefore he argued brilliantly in GAC proposals to ensure that the AEC would play a leading role in fundamental nuclear research. In one of his statements in support of nuclear research, he made what the then AEC Chairman, David Lilienthal, termed "as brilliant, lively, and accurate a statement as I believe I have ever heard."

In line with his case for AEC support of fundamental research, Oppie advocated that the Commission support such research in the universities and other research establishments; thus he helped initiate the incredible growth of science resulting from government–university cooperation.

Finally, regarding Oppenheimer's contributions on the GAC (and I have touched on only a few of them), he was a strong advocate of making fundamental scientific information available to all scientists and of distributing such materials as radioisotopes to scientists abroad, not only for medical investigation and therapy but for use in basic research.

Oppie's contributions to his government extended far beyond his service during the war and his work on the GAC.

While still at Los Alamos, he was one of the first to recognize that a nuclear test detection system should be established and so recommended when he was still with the Manhattan District. When he was chairman of the Committee on Atomic Energy of the Joint Research and Development Board, he was helpful on a number of occasions throughout the years 1948 to 1950 to the program conducting research and development on techniques for detecting nuclear explosions. I had the privilege of serving with him on the panel that evaluated and confirmed the report by early scientific detection experts that the Soviets had indeed broken the U.S. monopoly on nuclear weapons by testing a nuclear device of their

---

A strong advocate of making fundamental scientific information available to all scientists. . .

own on Aug. 29, 1949. He also served in a similar capacity to endorse the findings of the U. S. detection system in 1951 that the Soviets had conducted their second and third nuclear tests.

He served his government in numerous other capacities. He served in 1945 on Secretary Stimson's Scientific Panel of the War Department's Interim Committee and in 1946 on President Truman's Evaluation Committee for Operation Crossroads. He served the Joint Research and Development Board from 1947 to 1952 in many capacities, perhaps the chief of which was as member and chairman of its Committee on Atomic Energy. He was a member of the Naval Research Advisory Committee from 1949 to 1952 and of the Science Advisory Committee, Office of Defense Mobilization, from 1951 to 1954. He served on the Secretary of State's Panel on Disarmament in 1952 and 1953. This enumeration of his services to his country is only a representative fraction of the total of his contributions.

You are all familiar with Oppenheimer's leading role in formu-

A colloquium in Los Alamos after the war. Norris Bradbury, John H. Manley, Enrico Fermi, and Morris W. Kellogg in front row; Robert Oppenheimer and Richard P. Feynman in second row.

His many contributions to his country were recognized by three presidents. . .

---

lating the Acheson–Lilienthal Report of 1946, which called for the creation of an international authority to control all atomic energy work. Much of the substance of this plan, which emphasized the peaceful potential of atomic energy, was incorporated in the proposal later presented to the United Nations by Bernard Baruch. Although the Baruch proposal was rejected, it set the tone for future thinking in international control and cooperation and anticipated many of the subsequent ideas and much of what we hope to achieve in the future. Twenty years later our current hopes for a nonproliferation treaty owe much to this original groundwork. Oppenheimer's contributions to the Baruch plan were indicative of his farsightedness and depth of understanding as well as of his humanitarian outlook.

These are some of Robert Oppenheimer's contributions to his country in the area of governmental public service. But his public service contributions went far beyond this. He was tireless in his efforts to explain and interpret science—its meaning, its intellectual, cultural, humane, economic, political, and sociological implications—to the broadest possible audience. He did this by means of speeches to a diverse spectrum of audiences, by the written word in a wide range of publications, by participation before congressional committees, by numerous appearances on radio and television, and by active membership in many organizations and societies devoted to this cause.

His many contributions to his country were recognized by three presidents, who bestowed honors upon him. The first was President Truman, who in 1946 awarded him the Medal of Merit for his work at Los Alamos. The citation accompanying this award praised Oppenheimer for ". . .his great scientific experience and ability, his inexhaustible energy, his rare capacity as an organizer and executive, his initiative and resourcefulness, and his unswerving devotion to duty. . . ."

The second president to honor him was President Kennedy. This honor was in the form of an invitation to a White House dinner given in honor of Nobel Prize winners. President Kennedy had also decided, before his death, to present Oppenheimer with the Fermi Award.

J. Robert Oppenheimer receives the Enrico Fermi Award for the year 1963.

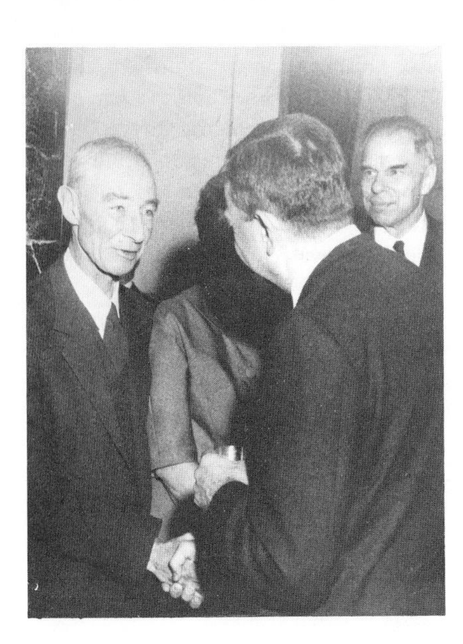

Edward Teller congratulating Robert Oppenheimer upon his receipt of the Enrico Fermi Award.

The third presidential honor he received was from President Johnson in 1963. On this occasion the President presented Oppenheimer with the AEC's Enrico Fermi Award. Making the award personally at the White House, President Johnson said, in part:

> Dr. Oppenheimer, I am pleased that you are here today to receive formal recognition for your many contributions to theoretical physics and to the advancement of science in our nation. Your leadership in the development of an outstanding school of theoretical physics in the United States and your contributions to our basic knowledge make your achievements unique in the scientific world.

It seems to me that President Johnson's reference to Oppenheimer's role as a teacher was singularly appropriate. His role as an extraordinary teacher must certainly be included among his public service and human contributions.

Robert Oppenheimer will go down in science history for his founding of what has been called "the American school of theoretical physics." In the first quarter of the twentieth century, theoretical physics had its home in the great laboratories and universities of Western Europe. After completing his undergraduate work at Harvard in 1925, Oppenheimer went to Europe to absorb all he could from these great centers of physics. He continued his studies at the Cavendish Laboratory at Cambridge where he became familiar with the work of Rutherford, Blackett, Dirac, and especially of Niels Bohr and Max Born.

Born was so impressed with Oppenheimer that he invited him to work with him at the University of Gottingen, and it was there that Oppie, at the ripe old age of 23, did his Ph. D. thesis on an application of Schrödinger theory to the photoelectric effect. His next two years were spent at Harvard and California Institute of Technology as a National Research Fellow and at Leyden and Zurich as a fellow of the International Education Board. While at Zurich he worked closely with Wolfgang Pauli and was greatly influenced by him.

Returning to the United States in 1929, Oppie accepted appointments as assistant professor of physics at both Berkeley and Cal Tech. Having gained a total command of the "new physics" of

---

Tireless in his efforts to explain and interpret science. . .

the era while in Europe, he now started, in his brilliant teaching, to shift the center of theoretical physics to the west coast of the United States. His list of outstanding students and associates reads like a who's who in physics. Among them were Willis Lamb, Phil Morrison, Hartland Snyder, Sid Dancoff, George Volkoff, Leonard Shiff, Bob Serber, Julian Schwinger, Bob Christy, and many, many others.

When Oppenheimer became Director of the Institute for Advanced Studies at Princeton in 1947, the physics department of the Institute became the new international mecca of theoretical physics, just as Copenhagen had served this role in the 1920s and 1930s. Pauli, Dirac, and Yukawa often came to Princeton during the Oppenheimer era at the Institute, and Gell–Mann, Goldberger, Chew, Low, Nambu, Dyson, Pais, Lee, and Yang were among the many who worked there under Oppie's inspired leadership.

Although his greatest contribution to science was probably in his role as an inspiring teacher, organizer, and catalyst of the new physics, Oppenheimer was a creative scientist who made many significant contributions to theoretical physics.

It is virtually impossible to summarize in the short time of this talk the additional "human" contributions of Robert Oppenheimer,

J. Robert Oppenheimer when he was Director of the Institute for Advanced Studies, Princeton University.

just as it is almost impossible to separate them from his scientific contributions. Those scientists who knew him well and worked with him closely were as impressed by the scope of his knowledge and interest in other matters—in languages, literature, the arts, music, and the social and political problems of the world—as they were by his scientific wisdom. Above all, those who knew him, read his writings, or heard him speak were impressed by his fervent desire to see and relate an order and purpose in the entire spectrum of human existence.

Oppenheimer was probably unique among scientists of our age in his effect on and high standing among other scientists. His magnetic—really electric—personality, his charismatic presence, and his unique style commanded attention in a manner equalled by

Upon Oppenheimer's last visit to the Los Alamos Laboratory in May 1964, he and Norris Bradbury toured the local science museum.

few scientists. His basically humanitarian outlook and his obvious concern for the overall welfare of humanity were widely recognized and appreciated throughout the world of science. And these qualities carried over to the world of nonscientists to an extent that was almost without parallel.

The passing of Robert Oppenheimer not only marks the passing of an era of physics but also portends an irreplaceable loss to the world of all scientists and nonscientists alike. ■

ATOMIC PIONEER AWARD

At the Atomic Pioneer Award ceremony: Glenn T. Seaborg, President Richard M. Nixon, General Leslie Groves, Vannevar Bush, James B. Conant.

Presentation of the Atomic Pioneer Award to Vannevar Bush, James B. Conant, and Leslie R. Groves, Feb. 27, 1970

■ **DR. SEABORG:** Mr. President, we are honoring three old friends here, friends of mine in the field of atomic energy and friends of yours.

You are presenting them with the Atomic Pioneer Award. This is the first of a kind and the only presentation that will be made of this award because there is only one Dr. Bush, only one Dr. Conant, and only one General Groves.

No one could be in their class with respect to the field we are honoring them for today.

I would like to begin by reading the citation for Dr. Vannevar Bush:

> For his exceptional contributions to the national security as Director of the Office of Scientific Research and Development in marshalling the resources of American science for national defense during World War II and for his pioneering leadership as a Presidential advisor in fostering the establishment of new federal agencies, including the National Science Foundation and the Atomic Energy Commission, which have made possible the unprecedented growth of scientific research and development in the last two decades.

The first of a kind and the only presentation that will be made of this award. . .

That is signed by President Nixon and the five Commissioners of the Atomic Energy Commission.

THE PRESIDENT: I would emphasize your comment that this is most unusual. The President has the responsibility of presenting many awards, the Medal of Freedom and others, and they are always to distinguished people. But this is the only award that I know of that is a one-time award and presented only to the three men who are here.

The award was created for these three pioneers in the field of atomic energy. I think, therefore, it has a unique quality that no other award we have ever presented has had.

We want to congratulate all of you.

DR. SEABORG: Now the award to Dr. James B. Conant; the citation reads:

> For his exceptional contributions to the national security as Chairman of the National Defense Research Committee in overseeing the successful development of weapons systems, including the atomic bomb, during World War II and for his pioneering leadership in the nation's atomic energy program after the war as Chairman of the Committee on Atomic Energy of the Joint Research and Development Board and as a member of General Advisory Committee to the Atomic Energy Commission.

THE PRESIDENT: You were also chairman of other commissions, Dr. Conant. You started it all.

DR. SEABORG: Now I will read the citation for Gen. Leslie R. Groves.

> For his exceptional contributions to the national security as Commanding General of the Manhattan Engineer District, United States Army, in developing the world's first nuclear weapons during World War II, and for his pioneering efforts in establishing administrative patterns adopted by the Atomic Energy Commission in effecting the use of atomic energy for military and peaceful purposes.

THE PRESIDENT: We have representatives of the Senate and House here. I wonder if they would like to say a word to our three award winners.

SENATOR PASTORE: I think that mankind owes these three gentlemen a tremendous debt of gratitude. If it had not been for the development of the bomb, I think we would not have been able to withhold the onslaught of Communism in the world.

This weapon was the mainstay in Europe. I think it is still a deterrent today and is really helping the security of this country and the free world. It all began with you. Without you, it would not have happened.

CONGRESSMAN HOSMER: Both for myself and for Congressman Holifield, who regrets very much that he could not be here today, I want to express deep appreciation, particularly because, although you started this work in a warlike fashion, today the emphasis is on what the atom can do for the world. All future

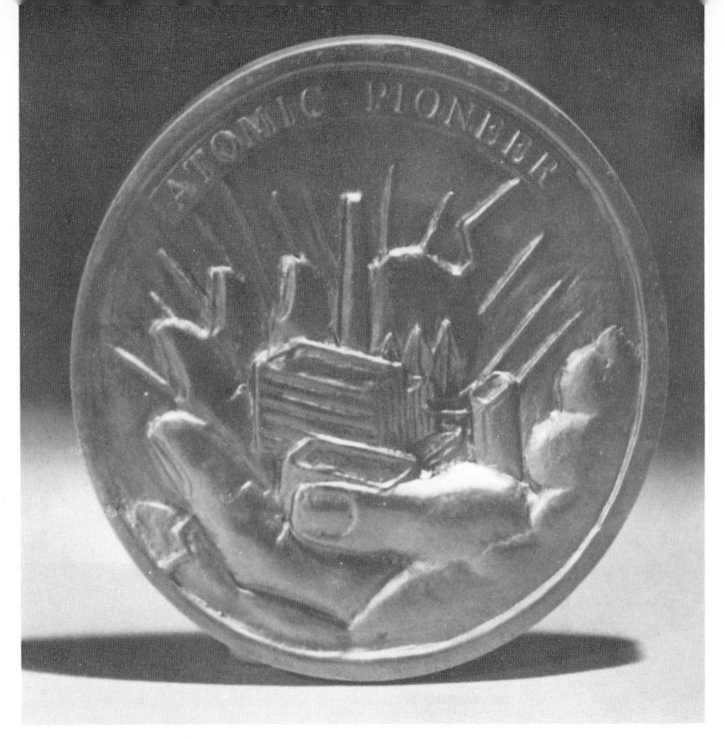

The Atomic Pioneer Award medal.

generations will owe you gentlemen a vast debt.

THE PRESIDENT: I think we can say, too, that this really quantum breakthrough in knowledge had a very dramatic effect on the thinking of the people not only of this country but also of the world, particularly the scientific community.

After this breakthrough came the breakthrough in space and everything else.

DR. SEABORG: I think we should emphasize the tremendous peacetime applications that we are reaping the benefits of now, and we can look forward to even greater benefits in the future. ■

Upon Presentation of the ENRICO FERMI AWARDS

The Enrico Fermi Award medal.

OTTO HAHN AND FRITZ STRASSMANN

Hofburg Palace, Vienna, Austria, Sept. 23, 1966

■ The presentation of the Enrico Fermi Award for 1966 here in the Festsaal of the Hofburg Palace is indeed a significant event in many respects. It is significant for the President of the United States and the Atomic Energy Commission because this is the first time the award has been presented outside the United States and to other than American citizens. It is also the first time that a woman has been named to receive the award. I know it is significant to many of the distinguished guests in our audience today because many of you have the honor and pleasure of knowing, and some of having worked with, the recipients of the award. It is personally a very significant moment for me to be making this presentation here today because of the direct influence the recipients' achievements have had on my work and my life. I am sure that to Otto Hahn, Lise Meitner, and Fritz Strassmann it is significant to have this well-deserved recognition of their achievements from the American people added to the accolades they have received from their fellow citizens and scientists around the world.

Let me, for a moment, elaborate on these points. With this year's presentation of the Fermi Award here in Vienna the award becomes, in effect, an international prize in nuclear science. I believe that is as it should be. I hardly need to tell this audience that we are striving to make the promises of the peaceful atom a reality for all men just as our knowledge of the atom is the heritage of all men. In this respect, it is also fitting that we have chosen to make this presentation in Vienna, the home of the International Atomic Energy Agency, an organization in which the United States and almost 100 other nations of the world have placed great faith and whose work we all firmly support.

We are honored today to have such a distinguished audience share this moment with Drs. Hahn and Strassmann. We know that you also send your good wishes to Dr. Meitner, who, unfortunately, cannot be here today but to whom we will present the award in England next month. In addition to knowing and working with Otto Hahn, Lise Meitner, and Fritz Strassmann, many of you here today also had the privilege and pleasure of knowing the man in whose honor this award is made, Enrico Fermi.

The presentation of the Enrico Fermi Award this year is unique in yet another respect. Seldom before has an audience been gathered which includes both the very founders of a branch of science and those who have so advanced the important work which that science has ushered in. I refer, of course, to nuclear science and to the fact that the Nuclear Age was born and nurtured through the work of so many present in this hall.

Finally, if you will allow me a personal note, the presentation of the Fermi Award today is a great and meaningful occasion for me. The work of Otto Hahn, Lise Meitner, and Fritz Strassmann was directly related to my early career, and their discoveries, needless to say, greatly affected my future. I can recall vividly how as a young graduate student at the University of California at Berkeley in the mid-1930s, and later in connection with my work on plutonium, I used Dr. Hahn's book *Applied Radiochemistry* almost as my bible. How well I remember the night in January 1939 when, at a seminar in the Department of Physics at Berkeley, I first learned of the results of Hahn and Strassmann's experiments in which they verified the fissioning of uranium! It is impossible to describe the excitement this news produced in me. I recall that after the seminar was over I walked the streets of Berkeley for hours turning over and over in my mind the import of that discovery. Its impact, of course, is a story known throughout the world. It led to the first observation of the practical nuclear fission chain reaction by Enrico Fermi and his coworkers at the University of Chicago in December 1942 and thus became the basis for all our nuclear science and technology today.

In 1955 I had the great privilege of introducing Dr. Hahn as a lecturer at Berkeley, and I recall how he intrigued and charmed his audience on that eventful occasion. Today it is an even greater privilege to be on the same platform with him and Dr. Strassmann for the purpose of honoring them before so many of their distinguished friends and colleagues.

It is a pleasure now for me to read the citations of the Enrico Fermi Award for 1966 to Otto Hahn and Fritz Strassmann:

> The President of the United States of America and the Atomic Energy Commission, pursuant to the authority of the Atomic Energy Act of 1954 for the granting of awards for especially meritorious contributions to the development, use or control of atomic energy, grant The Enrico Fermi

Award to Otto Hahn for pioneering research in the naturally occurring radioactivities and extensive experimental studies culminating in the discovery of fission.

July 29, 1966

Signed: Lyndon B. Johnson
Glenn T. Seaborg
Wilfrid E. Johnson
Samuel M. Nabrit
James T. Ramey
Gerald F. Tape

The President of the United States of America and the Atomic Energy Commission, pursuant to the authority of the Atomic Energy Act of 1954 for the granting of awards for especially meritorious contributions to the development, use or control of atomic energy, grant The Enrico Fermi Award to Fritz Strassmann for contributions to nuclear chemistry and extensive experimental studies culminating in the discovery of fission.

July 29, 1966

Signed: Lyndon B. Johnson
Glenn T. Seaborg
Wilfrid E. Johnson
Samuel M. Nabrit
James T. Ramey
Gerald F. Tape

Otto Hahn and Fritz Strassman receive the Enrico Fermi Award for 1966.

LISE MEITNER

Cavendish Laboratory, Cambridge, England, Oct. 23, 1966

■ I am pleased to be here today to honor a truly remarkable lady to whom I have just presented the Enrico Fermi Award for 1966, an award which she shares jointly with her colleagues Professors Otto Hahn and Fritz Strassmann.

Professor Meitner, who was born in Vienna on Nov. 7, 1878, has been one of the great pioneers in nuclear physics. Her first paper, published in 1906, demonstrated the scattering of alpha rays. In 1907 she went to the University of Berlin to conduct theoretical studies under Max Planck and to begin her 30-year collaboration with Otto Hahn.

In 1917 Professor Meitner was entrusted with the organization of a Department of Radioactivity at the Kaiser Wilhelm Institute. Lise Meitner and her associates discovered a number of radioactive isotopes, including AcC″ (thallium-207), ThC″ (thallium-208), and protactinium, the latter having been discovered in 1918 with Otto Hahn. The emission of the so-called Auger electrons was first described and correctly interpreted by Dr. Meitner, and in 1925 she showed that the electron "lines" were emitted after and not before the radioactive transformation.

Around 1930 she worked for several years with her own students on purely physical questions, such as the heat generated in beta decay, and confirmed that the continuous energy spectrum of the primary beta particles was not due to secondary energy losses. She also studied the anomalous absorption of gamma rays in heavy elements. The latter was eventually found to be caused by pair production. Professor Meitner was the first to observe pair production by gamma rays in a cloud chamber.

Her work with neutrons began in 1932 with disintegration experiments of light nuclei and continued later with heavy elements. It culminated in the experiments on irradiation of uranium and thorium with neutrons. In 1935 she persuaded Professor Hahn to join her in the study of the neutron-induced activity of uranium, discovered by Fermi and ascribed by him to transuranium elements. In 1938 she and her coworkers showed in the course of these experiments that the 25-minute uranium must be due to resonance capture in uranium-238.

Radium-like isotopes were also found among the products of neutron irradiation of uranium. In 1938, before the "mystery" of these isotopes was solved, Lise Meitner was, for political reasons, forced to leave Germany. She then settled in Stockholm and started work at the Nobel Institute for Physics. In December 1938 and January 1939, her work with Professor O. R. Frisch, explaining the fission process in heavy elements in terms of instability against deformation, played a most important role for later theoretical investigations. This work was described in a letter written Jan. 16, 1939, and published shortly thereafter, in which she and Frisch correctly surmised that the radioactive products produced through irradiation of the uranium by neutrons were due to the splitting of uranium, a process to which they applied the term "fission." They also correctly deduced the expected energy release, about 200 MeV, and that each fragment would "give rise to a chain of disintegrations."

Professor Meitner continued her work with investigations of the nature of various fission products, the problems of asymmetric fission, and various problems of gamma spectroscopy. In 1950, shortly after the creation of the shell model, she pointed out the applicability of this model for various fission problems, and, in an

Lise Meitner receives the Enrico Fermi Award for 1966.

Lise Meitner and Otto Hahn in their laboratory in the 1930s.

Leading to the discovery of fission. . .

article in 1952, she discussed interesting relations between thermal fission, fast fission, and similar nuclear processes and the shell model in connection with magic-number elements.

Although it was Joliot, Halban, and Kowarski who predicted the possibility of a chain reaction and proved that more neutrons are generated in the fission process than are absorbed, Lise Meitner, by her part in the initiation of the vital experiments in Berlin and in the evaluation and interpretation of the experimental results of Hahn and Strassmann, had a key role in demonstrating the possibility of gaining nuclear energy from the fission reaction.

These discoveries by Lise Meitner and her joint work with Hahn and Strassmann provided the base for Enrico Fermi's subsequent achievement of a sustained and controlled release of energy by nuclear fission. In the words President Johnson sent to the Vienna ceremony, "It was they who opened the doors to the atomic age," and "It was they who gave the present generation of scientists the greatest power ever known to man."

I think it is most fitting and appropriate that the woman who pioneered in these achievements should be the first woman to receive the Enrico Fermi Award and, along with Professors Hahn and

Strassmann, the first to receive the award on an international basis.

I would now like to read the words of the citation:

> The President of the United States of America and the Atomic Energy Commission, pursuant to the authority of the Atomic Energy Act of 1954 for the granting of awards for especially meritorious contributions to the development, use or control of atomic energy, grant the Enrico Fermi Award to Lise Meitner for pioneering research in the naturally occurring radioactivities and extensive experimental studies leading to the discovery of fission.
>
> July 29, 1966
>
> Signed: Lyndon B. Johnson
> Glenn T. Seaborg
> Wilfrid E. Johnson
> Samuel M. Nabrit
> James T. Ramey
> Gerald F. Tape

JOHN A. WHEELER

The White House, Washington D. C., Dec. 2, 1968

■ **THE PRESIDENT**: Chairman Seaborg, Dr. Hornig, Distinguished Ladies and Gentlemen:

Twenty-nine years ago, a scientific paper was published which bore a very simple title: "The Mechanism of Nuclear Fission." That paper became the cornerstone for all the later understanding in this field, and its publication was a step forward unlocking the fantastic secrets of the nuclear age.

Today we have come here to the historic East Room of the White House to honor the man who, with Niels Bohr, wrote that historic paper: Dr. John A. Wheeler of Princeton University, scientist, teacher, innovator, pioneer of modern physics, man of thought, and man of action.

For the average layman, merely to read the list of Dr. Wheeler's achievements is to realize how incredibly complicated this world in which we live has become. Most of us are not so surefooted as you are, Dr. Wheeler, in the complex world and in the difficult vocabulary of the nuclear scientist.

But there is one thing that all of us, laymen and scientists alike, can understand: It is the idea that the human mind must be free to range as far and as freely as it can, unfettered and unconstrained.

John A. Wheeler receives the Enrico Fermi Award for 1968.

You are one who has chosen, like Ulysses, "To follow knowledge like a sinking star, Beyond the utmost bound of human thought."

Our hope is to sustain and to support you in that voyage.

Today we honor a great scientist with the Enrico Fermi Award of the Atomic Energy Commission. In receiving this award, he joins such explorers of the scientific frontier as Dr. John von Neumann, Dr. Eugene Wigner, and that great and that good and that talented public servant, than whom there is no better, Dr. Glenn Seaborg, and Dr. Robert Oppenheimer.

In honoring Dr. Wheeler we honor in addition the idea of excellence, and we honor all who make the pursuit of knowledge their vocation.

Dr. Wheeler, it is a very great pleasure to me to welcome you and Mrs. Wheeler—and three other generations of Wheelers—here at the White House today. You do us honor by your visit.

You give your country great satisfaction and assurance.

DR. SEABORG: Mr. President, I would like to read the wording of the Enrico Fermi Award to John A. Wheeler:

> This award is for his pioneering contributions to understanding nuclear fission and to developing the technology of plutonium production reactors and his continuing broad contributions to nuclear science.

I might add, ladies and gentlemen, this award is signed by Lyndon B. Johnson, President of the United States and also by the five Atomic Energy Commissioners. ■

WALTER H. ZINN

Hilton Hotel, San Francisco, California, Dec. 2, 1969

■ I am pleased that we have such a large and distinguished gathering for this event. Perhaps this audience better than any other can appreciate the significance of the Enrico Fermi Award. By the same standard, you also can appreciate why this year's recipient, Dr. Walter H. Zinn, so well deserves the honor we are about to bestow on him.

This award is presented on the twenty-seventh anniversary of the first controlled nuclear chain reaction carried out by Enrico Fermi and his associates on Dec. 2, 1942. One of Enrico Fermi's foremost associates on that historic day in Chicago was our honored colleague Wally Zinn. But that was not Wally's first significant contribution in the nuclear field. He helped write the opening chapters of the nuclear age more than 30 years ago, and there are few who can equal his record of continuous service in the atomic energy field.

At Columbia University in the 1930s, Wally was associated with some of the earliest research in nuclear physics in this country. When the discovery of nuclear fission was announced in January 1939, he found himself on the leading edge of a new field of research with unprecedented opportunities. The American scientific community was electrified by the experimental evidence from Germany, and within a few days the fission process had been confirmed by

Walter Zinn receives the Enrico Fermi Award for 1969.

For the world's first reactors and contributions to atomic reactors for research, production, propulsion, and electric power. . .

Dr. John Dunning and his associates in the Pupin Physics Laboratory at Columbia, marking the first demonstration of fission in the Western Hemisphere.

Dr. Zinn had no direct part in the first splitting of the uranium atom in this country, but, because of his five years of experience on neutron investigations, he immediately became a valuable member of the nuclear physics team of Fermi, Leo Szilard, and others at Columbia who quickly embarked on the exploration of the nuclear fission process. On Mar. 3, 1939, Szilard and Zinn conducted an experiment at the Pupin Laboratory in which they observed the emission of neutrons in fission of uranium. The flashes of light they saw on a screen provided experimental evidence that a chain reaction was possible and that the vast energy released in the fission of uranium could be utilized either in an atomic power plant or in an atomic bomb.

As nuclear research disappeared behind the cloak of national security in the early 1940s, the Columbia group wrestled with the basic experiments that were to support the new structure of reactor technology. Their primary goal was to achieve a chain reaction, and, after conducting preliminary experiments at Columbia, the group moved to the University of Chicago in the spring of 1942. Dr. Zinn was placed in charge of construction of Chicago Pile No. 1 (CP-1), which became the world's first nuclear reactor.

Dr. Leona Woods (now Mrs. Willard Libby), the only woman scientist working on that secret project, relates that Wally would not let her take part in the actual construction of the pile. She says that working with graphite blocks was a dirty, dusty job, and all workers wore overalls, goggles, and masks. Everybody looked alike. Wally had Leona excluded from this part of the hard-driving operation because, if he had to say nasty words to somebody, he did not want it to be a girl.

By working around the clock, the Fermi team was able to complete assembly of the pile by the end of November. At the important criticality test on December 2, Wally handled the emergency control rod called "zip," which he pulled out manually in the startup operation and reinserted to terminate the chain reaction after 28 minutes of operation.

It was in those hectic days of 1942 in Chicago that I first came to know Wally. I had read with avid interest his earlier reports describing his research in nuclear physics, and I was pleased to meet him at the Metallurgical Laboratory. I recall with special gratitude the invitation that the Zinns extended to Mrs. Seaborg and me to have Christmas dinner with them in 1942. The warmth of the Zinn household and their sharing of a quiet family dinner with us meant a great deal to two Californians who might otherwise have felt homesick on a wintry December day; this act of friendship was characteristic of Wally Zinn.

During World War II Dr. Zinn made many important contributions to nuclear science and technology. He continued his fundamental research on fast neutrons, which contributed directly to the design of the first nuclear weapons. He assisted Fermi in basic design studies for the Hanford production reactors and rebuilt the original chain-reacting pile with some design improvements (the rebuilt pile was designated CP-2) at a new site near Chicago. In 1944 he also built the CP-3, the world's first heavy-water-moderated reactor, which provided the basic technology for the Savannah River production reactors.

By the end of World War II, Dr. Zinn had become one of the world's leading authorities on reactor physics and technology. Many

Scale model of the reactor used by Walter Zinn and his associates on Dec. 2, 1942, to achieve the first self-sustaining nuclear reaction.

Walter Zinn measures the energy of a beam of neutrons.

of the top scientists who had worked on the atomic bomb project during the war rightly wanted to return to the quiet of their university laboratories. Wally Zinn was one of those who accepted the challenge of the peaceful atom and stayed on the job. In fact, before the war was over, he had already developed the design concept for a fast breeder reactor.

Dr. Zinn was named Director of Argonne National Laboratory in 1946, and, in the early years of the Atomic Energy Commission, he was largely responsible for establishing the course and direction of reactor development in this country. He made the Argonne Laboratory a national and world center for reactor development, a fact which the Commission recognized in 1948 when it centered a substantial portion of AEC's reactor development activities at Argonne. As an advisor to the Commission, he drafted the AEC's first reactor development program in 1947 before the Division of Reactor Development was formally established in 1948. His proposal to build the first fast breeder reactor, originally called Zinn's Pile, was the first reactor project to be approved by the Commission.

Completion of what became known as Experimental Breeder Reactor No. 1 (EBR-1) was undoubtedly one of the high points of Dr. Zinn's distinguished career. He spent from May to December 1951 at the National Reactor Testing Station in Idaho personally supervising the critical experiments and early tests. By using three different types of fuel rods, which he characterized as "standard lean," "standard fat," and "long fat," he found the combination that would produce an even distribution of neutrons through the core, and he brought the reactor to criticality on August 24. After low-power testing, the reactor generated the world's first useable electricity from nuclear power on Dec. 20, 1951; this was another first for Wally Zinn.

The next day all the electrical power in the reactor building was

On Aug. 26, 1966, President Johnson, with members of the Atomic Energy Commission, unveiled a plaque that designated the Experimental Breeder Reactor-I (EBR-I) as a National Historic Landmark to celebrate the notable Dec. 20, 1951, event when the EBR-I was the first nuclear reactor to generate useful amounts of electricity. Left to right: Gerald F. Tape, James T. Ramey, Glenn T. Seaborg, President Johnson, and Samuel M. Nabrit.

supplied by nuclear energy. Wally's communication of this news to Washington was a real peacetime Christmas present to the Atomic Energy Commission, which was then preoccupied with national defense.

While he was director of Argonne Laboratory, Dr. Zinn was also personally responsible for developing the design concept for the pressurized-water reactor, based on the original ideas of Alvin Weinberg, Eugene Wigner, and others. This concept was first used in the Mark I prototype, then in subsequent plants for nuclear submarine propulsion, which is so important to our national defense, and later for a whole generation of civilian nuclear power reactors. Building on these early studies, Dr. Zinn proposed the boiling-water-reactor concept and directed the design, construction, and initial operation of the boiling-water-reactor experiments. In these two broad areas, the pressurized- and the boiling-water-reactor concepts, Wally Zinn again played a key role in establishing the basis for an important part of present-day nuclear power reactor technology.

I remember being at the National Reactor Testing Station in Idaho in 1954 soon after the time the core of BORAX-1 was deliberately destroyed in a safety experiment. Wally Zinn was my gracious host on that occasion and, typically, took me around on a personal tour to show me the effects of the destructive test of that reactor.

After the successful operation of the Experimental Boiling Water Reactor in 1956, Dr. Zinn resigned as director of Argonne to set up his own engineering firm. Let me summarize briefly at this point the early contributions this man made in reactor development:

1. He was in charge of construction of the first reactor, the graphite-moderated CP-1.

2. He was responsible for the design and construction of the first heavy-water-moderator reactor, the CP-3, which established an important nuclear power reactor concept.

3. He launched early work on the pressurized-water- and boiling-water-reactor concepts, which are now the primary bases for nuclear power production throughout this country.

4. He built the first breeder reactor, a fast breeder cooled with liquid metal, which produced the first electricity by nuclear means, demonstrated the principle of breeding, and is now considered the highest priority breeder power reactor concept of the future.

It is no exaggeration to state that the nuclear reactors Dr. Zinn developed laid the foundation for nuclear power in this country and throughout the world. When you consider that nuclear power generation is now and may be in the foreseeable future the greatest peaceful application of nuclear energy, you gain the full impact of his tremendous contributions to nuclear science and technology.

Over the last 13 years as head of General Nuclear Engineering Corporation and later as a vice president of Combustion Engineering, he has continued to make important contributions to power reactor development and remains as one of the world's leading authorities in that field.

It is also appropriate that we honor him in the context of this annual meeting of the American Nuclear Society and the Atomic Industrial Forum because Wally Zinn was the first president of the American Nuclear Society in 1955.

I will now read the award citation:

> For his pioneering work in atomic energy, including the world's first reactors and the fast breeder reactor, and for his distinguished record of leadership and contributions to the development of atomic reactors for research, production, propulsion, and electric power.

Therefore, Wally, on behalf of President Nixon, the Atomic Energy Commission, and the people of the United States, I am pleased to present to you the Enrico Fermi Award for 1969. ■

NORRIS E. BRADBURY

Los Alamos Scientific Laboratory, Los Alamos, New Mexico, Aug. 29, 1970

■ The man responsible for the "Bradbury Years" at Los Alamos is, of course, Dr. Norris E. Bradbury. During these years we have gleaned many nuclear benefits through work at LASL. Such progress is not made without competent leadership and dedicated personal effort. These qualities I attribute to Dr. Bradbury. The laboratory has fared extremely well under his managership, and I have great admiration for his attention and devotion to LASL.

I suspect there were times when such devotion to his work conflicted with family obligations. And, I suspect, because of this, Dr. Bradbury over these years might have gotten into no small amount of trouble with his wife, Lois. But developing teamwork has always been one of Dr. Bradbury's great skills as a manager, and I think this attribute has enabled him to build many a bridge over troubled waters. Into the equation of this work, he has always been able to insert two factors, a devoted family and a

Dr. Norris Bradbury receives the Enrico Fermi Award for 1970.

devoted laboratory staff. He and his family and he and the members of the laboratory staff have compromised in many ways over his love affair with LASL through the good times and the bad. Teamwork, team research, and total involvement have always been prerequisites to a long life at Los Alamos, and Dr. Bradbury is a living example of this fact.

For 25 years at Los Alamos, Dr. Bradbury has kept his outlook in resonance with the changing times, with new research, and with the growth of the laboratory. I might say that for 25 years he has kept the windows of his mind open to the wind of change. And because he is this type of man, he has continued to see Los Alamos with new perspective and in a new light. But the retirement of Norris Bradbury on September 1 does not end his years of influencing the fortunes of Los Alamos. If I know Norris Bradbury, he is not yet finished reshaping the character of this outstanding laboratory and wonderful community. If one should ask him about his future relation to Los Alamos, I think Norris would answer like the old Vermont farmer who, when asked whether he had lived in his village all his life, answered, "Not yit."

Today we are here to recognize Dr. Bradbury's contributions to the Los Alamos Laboratory, to the nuclear community, and to society. The U. S. Atomic Energy Commission has selected Dr. Bradbury to receive the Commission's Enrico Fermi Award for 1970. Dr. Bradbury is the fourteenth recipient of the award, which is presented annually for outstanding scientific achievements or contributions to engineering and technical management in the development of atomic energy.

The award is, of course, named after the late Enrico Fermi, who headed the effort that resulted in the first sustained controlled nuclear chain reaction.

The AEC selected Dr. Bradbury for the award after consideration of nominations and recommendation by its General Advisory Committee. President Nixon approved the selection.

Dr. Bradbury is, as many here today know him to be, a modest man who has performed his task quietly and diligently without self-acclaim. It is time we took note of this man; our acclaim of him is long overdue.

Dr. Bradbury's association with the atomic energy program began in 1944, about two years after Enrico Fermi's first controlled reaction experiment. As a naval officer and professor of physics on leave from Stanford University, he was assigned to Los Alamos and placed in charge of the assembly of nonnuclear components for the first nuclear device, which was tested in July 1945 at Alamogordo.

In October 1945 he succeeded Dr. J. Robert Oppenheimer as laboratory director. Under Dr. Bradbury's leadership, LASL grew from a complex of temporary wartime structures to one of the nation's largest institutions dedicated to basic and applied research in virtually all fields of nuclear science.

As laboratory director Dr. Bradbury played a major role in planning nuclear weapons development, directing research on advanced weapon concepts, and conducting field tests. The U. S. tests from 1948 to 1952 in Eniwetok and Nevada resulted in data that completely revolutionized nuclear weapon technology and served as the scientific basis that has made the U. S. nuclear capability the cornerstone of the free world's security.

An outstanding physicist in his own right, Dr. Bradbury's

For leadership and direction of the Los Alamos Scientific Laboratory and contributions to national security and atomic energy. . .

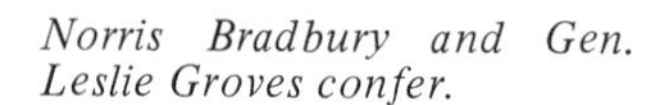

Norris Bradbury and Gen. Leslie Groves confer.

incisive technological vision saw great promise in the peaceful uses of nuclear energy. While maintaining the laboratory's defense activities, he directed LASL in becoming a primary center for research in the chemistry and metallurgy of uranium and plutonium. Under his leadership the laboratory also became the leader in nuclear rocket propulsion technology and space nuclear systems and a pioneer in the development efforts on controlled thermonuclear reactions. He promoted major activities in nuclear safety and techniques to enhance our capability to safeguard nuclear material.

Typical of Dr. Bradbury's foresight to achieve the necessary balance between peaceful and defense efforts, he established the laboratory as one of the centers for unclassified research in medium-energy physics. The Los Alamos Meson Physics Facility, which is nearing completion, will serve scientists and students from the regional and national academic community. He has also promoted and directed various biomedical programs that have enlarged our fundamental knowledge of the consequences and limiting values of radiation exposure to man. All these accomplishments have been achieved here at the Los Alamos Scientific Laboratory under Dr. Bradbury's direction.

We have cited just a few of Dr. Bradbury's many personal accomplishments. We are very honored to be here today to pay tribute to him and his work. In honor of this occasion I am pleased to read a special message from President Nixon.

Dear Dr. Bradbury:

Your brilliant and enduring contributions to the field

At the nuclear-rocket reactor test site.

of nuclear energy have earned you the pride of all your fellow citizens and the admiration of countless others throughout the world. Nothing could please me more than to congratulate you on receiving one of our Government's highest honors: The Enrico Fermi Award.

The finest reward for your accomplishments must surely be the knowledge that your persevering talents have so immeasurably speeded human progress and uplifted the quality of life. There is little I can add to this distinction but my strong assurance that this Administration will remain totally committed to the noblest goal of your successful career: to direct our nuclear energy toward the benefit of all mankind.

I know that I am joined by countless colleagues and admirers in the hope that your retirement years may be filled with all the contentment you have earned by your selfless public service.

Sincerely,

Richard Nixon

Now I am pleased to present to you the Enrico Fermi Award for 1970. I will read the award citation:

For his inspiring leadership and superb direction of the Los Alamos Scientific Laboratory throughout one quarter of a century, and for his great contributions to the national security and to the peacetime applications of atomic energy.

Others who have received the Enrico Fermi Award. . .

John von Neumann *(1956)*

Eugene P. Wigner *(1958)*

Glenn T. Seaborg *(1959)*

Hans A. Bethe *(1961)*

Edward Teller *(1962)*

Admiral H. G. Rickover *(1964)*

At the Atomic Energy Commission's twenty-fifth anniversary celebration: (left to right) Glenn T. Seaborg, Wilfrid E. Johnson, Rep. Melvin Price, M. Sterling Cole, Lewis L. Strauss, Mrs. Gordon Dean, David E. Lilienthal, Edward E. David, Sen. John O. Pastore, Hon. Bourke Hickenlooper, Sen. Clinton P. Anderson, James T. Ramey, and Rep. Chet Holifield.

PART TWO

Historic Landmarks

NATIONAL MANDATE FOR ATOMIC ENERGY

A 25-YEAR REVIEW

Twenty-fifth Anniversary of the AEC, Aug. 1, 1971

■ Twenty-five years ago this morning seven members of the Special Senate Committee on Atomic Energy gathered around President Truman's desk at the White House as he affixed his signature to Senate Bill 1717, which thereby became the Atomic Energy Act of 1946.

The bill President Truman signed that morning a quarter of a century ago was no ordinary piece of legislation. It had emerged from months of congressional and public debate over the life-and-death issues the dramatic advent of atomic energy had revealed to the world at Hiroshima. The new Act, as unusual in its provisions as the events which produced it, reflected a fundamental reexamination by the Congress of the national predicament at the end of World War II. In terse terms of understatement, the Act described the looming presence of atomic energy in the world of that day:

> Research and experimentation in the field of nuclear chain reaction have attained the stage at which the release of atomic energy on a large scale is practical. The significance of the atomic bomb for military purposes is evident.

If these facts were clear to everyone, the future was not. The Act went on to state:

> The effect of the use of atomic energy for civilian purposes upon the social, economic, and political structures of today cannot now be determined .... It is reasonable to anticipate, however, that tapping this new source of energy will cause profound changes in our present way of life.

I am sure that to some of us, with 25 years of hindsight, these

Members of the Special Senate Committee: (left to right) Senators Tom Connally, Eugene D. Millikin, Edwin C. Johnson, Thomas C. Hart, Brien McMahon, Warren R. Austin, and Richard B. Russell, on Aug. 1, 1946, gathered around President Truman as he signed the bill which thereby became the Atomic Energy Act of 1946 and, as part of it, established the AEC.

words sound overblown and exaggerated. The world has changed dramatically since 1946, and yet it is not clearly evident that nuclear energy has been the prime mover of our generation. For the general public such things as television, shopping centers, interstate highways, and air conditioning may have seemed to be more obvious sources of change, but the obvious does not always have the greatest significance.

Let us read a little further in the 1946 Act. It established the Atomic Energy Commission, itself a unique institution in governmental organization, and assigned to the Commission an exceptional responsibility. In the words of the Act, the policy of the United States was that

> . . . subject at all times to the paramount objective of assuring the common defense and security, the development and utilization of atomic energy shall, so far as practicable, be directed toward improving the public welfare, increasing the standard of living, strengthening free competition in private enterprise, and promoting world peace.

What can we say today about our success in achieving these five goals?

Overshadowing all the others in the language of the Act was that of assuring the common defense and security. For a few months after the end of World War II, there were some hopes that the military aspects of nuclear energy could be subordinated to the peaceful aspects and that, under the system of international controls, this new energy source could be exploited for peaceful purposes. The somber events of 1948 and 1949, however, demonstrated the accuracy of the original assessment by those who drafted the 1946 Act. Nuclear weapons were to become the keystone of our national defense system, and the Atomic Energy Commission under the Act was called upon to accept the major responsibility for designing and building a new arsenal of weapons unprecedented in both their technical sophistication and their destructive power.

The existence of a stockpile of nuclear weapons has never been a source of comfort even to those who have the greatest confidence in our ability to carry out the responsibilities it imposes upon us. In recent years the public press has had much more to say about the dangers of the nuclear stockpile than about its benefits. Yet in taking a broad historical view (as seems appropriate on a silver anniversary), we must accept the fact that the terrifying prophesies of the past quarter century have not come true. The world has so far avoided nuclear war, and students of current affairs can cite instances in which the nuclear stockpile has helped to preserve the peace. Many of us who helped develop nuclear weapons during World War II profoundly hoped that nuclear warfare would prove too horrible for any rational nation to contemplate. Twenty-five years later we can see some evidence that this hope may be fulfilled.

I think it is, therefore, reasonable to state (and certainly without any sense of overweening pride or joy) that the Commission has been successful in carrying out its mandate to assure the common defense and security. Some of the best scientific talent in the nation has gone into this endeavor. Strictly as research institutions our weapon laboratories at Los Alamos and Livermore rank among the best in the world, representing an investment in talent and resources which has proved well worth the cost. They also clearly demonstrate a point I have often made in recent years—that, through skillful use of their talents and scientific knowledge, men can reap the promise of a new technology without necessarily incurring the hazards involved.

Beneath this umbrella of responsibility for national security, the Commission attempted to carry out the other four mandates of the

Act. In terms of improving the national welfare, I think first of the startling advances during the last quarter century in education in the nuclear sciences. This advance probably comes to my mind first of all because I recall so vividly the precarious state of this new research field at the close of World War II. Those of us who had been privileged to explore this new realm of science during the war were gravely concerned about the prospects for the future. Research in the nuclear sciences would require equipment and financial support on an unprecedented scale, far beyond the capabilities of traditional sources in universities and private research institutions. Under the 1946 Act the Commission had a leading role in creating new administrative machinery for federal support of research. The Commission also deserves credit for bringing into reality the new concept of the national laboratory. The research contract and the national laboratory became the key instruments for a system of research on a national scale which has helped to bring the nuclear sciences in this country to world preeminence. The impact of this system has extended far beyond the nuclear sciences to other fields of research and to American education in general. The Commission program has established new standards that have come to be accepted as the norm in educational institutions.

Under the topic of public welfare, I must also mention the extraordinary proliferation of the use of radioactive isotopes for industrial, medical, agricultural, and space applications. In the early years of the Commission's existence, this subject was perhaps overworked in citing the benefits of nuclear energy, probably because the advances of radioisotopes were immediately evident. In the absence of other concrete examples, it became customary to talk about radioisotopes. Now we take them much more for granted and can call attention to other kinds of accomplishments. But, again in the historical perspective, I think we should not forget that the use of reactor-produced radioisotopes began only 25 years ago tomorrow, when 1 millicurie of carbon-14 was delivered at Oak Ridge to the representative of a St. Louis hospital. Since that day both the variety and the quantity of available isotopes have increased to levels undreamed of a quarter of a century ago, with untold benefits in the diagnosis and treatment of diseases, in improving industrial operations, in promoting advances in agriculture, and in generating power in remote locations on the moon and in deep space.

There is no question in my mind that the public welfare has benefited in the last 25 years from the knowledge of the physical world which research in the nuclear sciences has made possible. The magnitude of this accomplishment is perhaps beyond the grasp of

At the Anniversary Commemoration. . .

Left to right: Lewis L. Strauss, Glenn T. Seaborg, Adm. H. G. Rickover, Edward Teller, Eugene P. Wigner, and Rep. Chet Holifield.

Left to right: Rep. Chet Holifield, Rep. Melvin Price, Sen. John O. Pastore, and Lewis L. Strauss.

Left to right: Lewis L. Strauss, Clarence E. Larson, and Edward E. David, Science Advisor to the President.

Left to right: Glenn T. Seaborg, Lewis L. Strauss, Edward Teller, and Eugene P. Wigner.

Left to right: David E. Lilienthal, Glenn T. Seaborg, and Lewis L. Strauss.

Commemorative plaques to. . .

David E. Lilienthal, the first Atomic Energy Commission Chairman. Presented by Chairman Glenn T. Seaborg.

Lewis L. Strauss, former Commission Chairman and a member of the first Commission. Presented by Glenn T. Seaborg.

Charter members of the Joint Committee on Atomic Energy: Rep. Melvin Price, the present Joint Committee Vice-Chairman; and Rep. Chet Holifield, former Joint Committee Chairman. Presented by Commissioner James T. Ramey.

Sen. Clinton P. Anderson, charter member of the Joint Committee, also former Joint Committee Chairman. Presented by Commissioner Clarence E. Larson.

Chairman Glenn T. Seaborg also presented commemorative plaques to Sen. John O. Pastore, Joint Committee Chairman, and Mrs. Gordon Dean, widow of former Commission Chairman Gordon L. Dean.

Commissioner Wilfrid E. Johnson presented plaques to Mrs. Richard Lane, daughter of the late Sen. Brien McMahon, and to Hon. Bourke Hickenlooper, former Joint Committee Chairman.

Commissioner Clarence E. Larson presented a plaque to Sterling Cole, former Joint Committee Chairman.

Plaques were sent to John A. McCone, former Commission Chairman, and Carl T. Durham, former Joint Committee Chairman.

And a special medal commemorating the Anniversary to President Truman at Independence, Missouri. . .

many Americans because its concrete results are indirect. In answer to the occasional cries from the wilderness that man is probing too deeply into the secrets of the universe, I reply that we can never know too much about the world around us. It is not knowledge itself but the use we make of knowledge that brings trouble. Our atomic energy program in the last 25 years has produced a veritable revolution in our understanding of both the physical and the biological sciences. Those who do not believe that should glance through some of the scientific journals of 1946. They reflect a world that seems incredibly remote today, not only in terms of the state of knowledge but also in the sense of our understanding of man's relation to his environment.

The 1946 Act also declared that atomic energy should be developed to improve the standard of living and to strengthen free competition in private enterprise. These words perhaps led some in the immediate postwar period to believe that a day of cheap and abundant nuclear power was about to dawn. I remember that those of us who served with Robert Oppenheimer on the Commission's first General Advisory Committee were worried about this over-optimistic expectation in 1947. Nuclear power has become a reality within a time frame much closer to the Committee's predictions than to the public's expectations. In fact, we are today just reaching the point where private industry is beginning to invest substantial amounts of capital in nuclear power plants on a strictly economic basis. Nuclear power has scarcely begun to have an impact on the world's standard of living, but, as we sharpen our predictions of future power needs, it seems inescapably certain that nuclear power will have a profound effect on the future standard of living. In this sense, the substantial investment of talent, resources, and money in nuclear reactor technology in the past quarter century has been an investment in the future. As the Commission stated in its report to President Kennedy in 1962, in 15 years we had reached the point at which nuclear power was becoming economically competitive. The technology of reactors using water as a moderator and coolant is now approaching maturity. Under the Commission's leadership the nation has now accepted a new commitment for the future as we seek to develop an economical fast-breeder reactor. Not only President Nixon and his administration but also the Joint Committee on Atomic Energy have provided the support this important effort requires.

"Strengthening free competition in private enterprise" conjures up the fears of 1946 that this new and little-understood technology of nuclear energy would fall into the clutches of a few giant monopolistic corporations. The 1946 Act, by placing complete control of atomic energy in the hands of the Commission, sought to avoid that disaster. From the beginning the Commission took this mandate seriously, and there has been a constant effort over the years to bring about an orderly transformation from complete government monopoly to a nuclear industry fully integrated into the national economy. This process began slowly, but by 1952 it was possible to begin thinking of a dramatic step toward industrial participation. What started as a piecemeal effort to amend the 1946 Act ended in a sweeping revision of the statute, which became known as the Atomic Energy Act of 1954. Under the new Act American industry could begin to play a major role in developing nuclear power, and the utilization of nuclear power for peaceful purposes could begin on an international scale. Ten years later, in 1964, the Congress, under the leadership of the Joint Committee, further amended the Act to permit private industry to own special nuclear materials as well as the plants in which they are used or produced.

So rapidly has the atomic energy industry grown in recent years that the government monopoly of 25 years ago seems to have faded into the remote reaches of history. It would seem that the architects of the 1946 Act achieved the goal they sought in assuring that this new form of energy would be developed for the benefit of all the people of this nation. Today we can broaden their intention and express the hope that nuclear energy can become the servant of all men everywhere.

The last mandate enjoined upon us by the 1946 Act was to use atomic energy to promote world peace. These words express one of man's highest and most elusive aspirations. As I suggested earlier, it is possible that nuclear energy in its military applications has prevented a global war, but it would be less than the truth to say that nuclear energy has brought us closer to world peace. I would hazard the prediction that nuclear weapons will never have much more than a passive role in the quest for world peace. As in the past, they may help to protect us while we work to build the kind of world that can exist in peace. But the kind of peace that all men seek will come only through the solution of the enormous social and economic problems that plague us today. I think it is no exaggeration to say that the development of nuclear energy in the last 25 years has opened up new opportunities for finding solutions to these problems. With patience, hard work, and clear thinking, we will, by Aug. 1, 1996, be closer to our ultimate goal as a result of the forces set in motion 25 years ago today. ■

The X-10 Reactor—

A Landmark of the Nuclear Age

Designation of the X-10 Reactor as a National Historic Landmark, Oak Ridge, Tennessee, Sept. 13, 1966

■ It is indeed a pleasure and a privilege for me to accept this plaque and certificate on behalf of the U. S. Atomic Energy Commission.

The designation "National Historic Landmark" is most certainly a coveted recognition, and the Commission is deeply appreciative of this honor bestowed by the National Park Service and the U. S. Department of the Interior.

I have had the pleasure of being present this year at the naming of three places involved in nuclear energy work as National Historic Landmarks: Room 307 Gilman Hall at the University of California at Berkeley, where plutonium was discovered; the Experimental Breeder Reactor No. I at the National Reactor Testing Station in Idaho, where electricity was first generated by nuclear power; and here at the X-10 Reactor, the world's first production reactor and the first to produce usable amounts of radioisotopes.

To have places and facilities associated with nuclear advances, particularly those with which one has had some personal affiliation, become recognized as National Historic Landmarks gives one a tremendous sense of satisfaction. I am sure that many of you here today share this feeling as you see the Nuclear Age, whose birth and early development you tended, progress and take on a more positive outlook for humanity. I believe that as the peaceful atom grows in its many applications and as it someday far overshadows the military application, these landmarks will hold an even more important place in our nation's history.

To be named a National Historic Landmark is particularly fitting for this the oldest of production reactors, which, in retirement, stands as a tribute to the inventiveness and technical expertise of nuclear scientists.

t the dedication: Alvin M. Weinberg, Director of the Oak idge National Laboratory, and Glenn T. Seaborg, Chairman f the Atomic Energy Commission.

The X-10 Graphite Reactor under construction in 1943. High-purity graphite for the reactor was machined in the building at right. In the radiochemical pilot plant at left, the bismuth–phosphate process would be "hot" tested on a semi-works scale with the highly radioactive plutonium–uranium slugs from the X-10 reactor.

At the fuel-loading face of the X-10 reactor. After the fuel slugs have been irradiated, they are pushed out of the reactor fuel channels, as shown, and dropped into a shielding canal at the back face of the reactor.

The X-10 Reactor was brought to life Nov. 4, 1943. . .

Loading fresh uranium fuel slugs into the reactor.

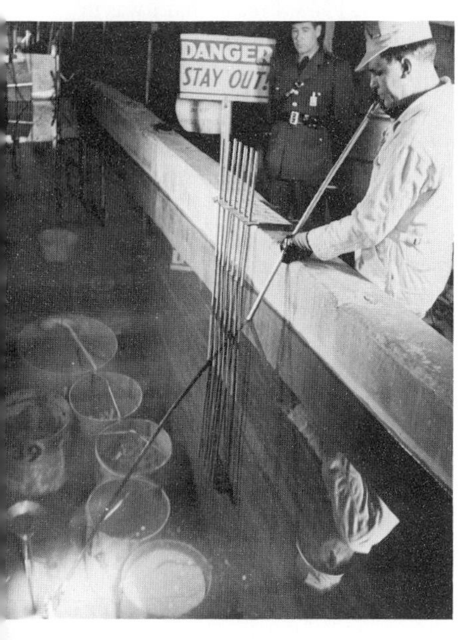

Worker at the X-10 reactor, with security guard watching, transfers the reactor-irradiated fuel slugs underwater in preparation for chemically separating the fissionable plutonium material from the irradiated slugs. The water shields the worker from the intense radioactivity but, being transparent, does not interfere with his view.

It is difficult to believe that almost 23 years have passed since the X-10 Reactor was brought to life, Nov. 4, 1943. I recall that I was in Oak Ridge at the time, and I know that many of you who are present today were also here and were deeply involved in those earliest days of the Nuclear Age.

Overshadowing the sheer excitement of having the reactor "go critical" was the stark realization of the profound importance of the X-10 project toward the war effort and the awesome stakes contingent on success or failure.

Of course the complete success of the initial mission of the reactor, and thus of the Manhattan Engineer Project in bringing World War II to a close, is now a dramatic chapter in American history.

I daresay that on Nov. 4, 1943, few of us present here in Oak Ridge at the birth of this reactor envisioned the rapidity with which the nation's program of peaceful applications of atomic energy would develop and the key role the X-10 Reactor would play in this important endeavor.

As most of you know, the reactor's initial function was to produce sufficient quantities of plutonium-239 to permit the

The first commercial shipment of radioisotopes on Aug. 2, 1946. . .

---

refinement of a chemical process for large-scale separation of this fissionable element.

This was a pilot-plant operation, and later the approach was scaled up and applied at Hanford, Washington, following closely the original Oak Ridge design and the work done at the Metallurgical Laboratory in Chicago.

Shortly after the end of the war, the X-10 Reactor began large-scale production of radioisotopes, and, with the first commercial shipment on Aug. 2, 1946, of 1 millicurie of carbon-14, a new era of the peaceful application of atomic energy was ushered in.

A World War II view of the Y-12 Plant at Oak Ridge, Tennessee, where uranium-235 was first separated on a large scale, in the electromagnetic separation plant. This plant was later dismantled and replaced by the present method of separation by gaseous diffusion.

The Electromagnetic Separation Plant. . .

The plant had five of these basic units, called "racetracks" for obvious reasons. These massive, elliptical, steel structures were each 122 feet long, 77 feet wide, and 15 feet high. The protruding ribs at the top are the silver-wound magnet coils. The boxlike cover that runs around the top of the racetrack contains a solid-silver bus bar roughly a square foot in cross section. Because of the scarcity of copper, about 14,700 tons of silver—worth $400 million— had to be used.

Experimental racetrack at Y-12 built in the summer of 1943 to test the alpha tracks for the electromagnetic process and to train operators. Workmen are preparing a vacuum tank for insertion between the magnet coils. Each racetrack contained 96 such vacuum tanks.

All the internals of the vacuum tank—the sources, collector, and liner—were fastened to a huge vacuum-tight door, or metal plate, which was installed in the tank as one unit by a special lift as shown.

Control panels for each tank were manned by hundreds of young women from the surrounding Tennessee countryside.

Oak Ridgers during the war years. . .

Hutment housing in early Oak Ridge.

Small hutments provided laundry and cleaning facilities to Oak Ridgers in 1945.

The main Oak Ridge shopping area, constructed by the army. The administrative buildings of the Manhattan Engineer District are visible in the background.

Oak Ridgers celebrate the end of World War II.

They pioneered in the gaseous diffusion method. . .

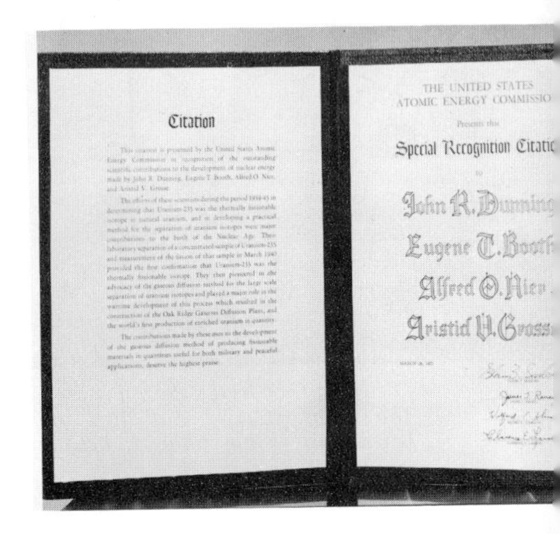

On Mar. 26, 1971, the United States Atomic Energy Commission presented (left to right) Eugene T. Booth, Alfred O. Nier, John R. Dunning, and Aristid V. Grosse with this citation for their "outstanding scientific contributions to the development of nuclear energy."

A wartime view of the Gaseous Diffusion Plant in Oak Ridge. This first of several diffusion plants is a huge U-shaped structure. Each side of the building is 2450 feet long, and the sides average 400 feet in width and 60 feet in height. The building covers 44 acres.

For years this reactor was the world's leading producer of radioisotopes, and as late as 1960 it was the major source of supply in the United States. Its isotope-production role decreased as private industry assumed this burden.

As a training instrument for reactor operators and technicians, the reactor lays claim to worldwide alumni, and the experiments and irradiations hosted within its core over the years are legion.

After operating efficiently and faithfully for 20 years, the X-10 Reactor was placed on standby, and its control rods were inserted for the last time in 1963. Alvin Weinberg, Eugene Wigner, and I were present at its "retirement ceremony" on Nov. 4, 1963, as were many of you who are here today. On that occasion Richard Doan gave what we might call the X-10's "farewell address."

In a sense the X-10 Reactor was the victim of its own success. The more efficient reactor systems at this laboratory and elsewhere which assumed the heavy work load the X-10 relinquished were the direct outgrowth of technology developed by this now-famous reactor. Even the modern nuclear power plants of today have an ancestry traceable to the X-10 pile.

From its secret birth to its reluctant retirement, the graphite reactor contributed immeasurably to the defense of this nation's liberty and to the betterment of man's everyday life.

Most assuredly the X-10 Reactor occupies a unique niche in the annals of American history and has more than earned its merit as a National Historic Landmark. ■

LARGE-SCALE ALCHEMY

Twenty-Fifth Anniversary at Hanford-Richland

At the Hanford twenty-fifth anniversary celebration (left to right): George Watt, Glenn T. Seaborg, Leslie R. Groves, and F. T. Matthias.

I am pleased to be sharing the platform with Gen. Leslie R. Groves. . .

Celebration Commemorating 25 years of progress at the Hanford Works and the City of Richland, Richland, Washington, June 7, 1968

■ Although this is Hanford's twenty-fifth anniversary year, it is hard to peg the birth of this great complex to any one day or week. In looking over the record, I noted that today, June 7, was the day in 1943 that construction was begun on the first Hanford pile; so I believe we have chosen an appropriate time for this celebration.

My own association with the site of the Hanford Engineer Works—"Site W" as it was called at that time—dates back to May 1944.

One of the considerations for choosing this site for the Hanford operation was its remoteness from centers of population. My first impression of the area led me to agree with the selection on this account. I recall looking over some of the flattest, most lonesome territory I had ever seen. Of course, there were other important factors involved in the selection. There was the mighty Columbia River to provide the necessary amounts of low-temperature water to cool the chain-reacting piles, and flanking the area at convenient distances were the Grand Coulee and Bonneville power networks.

The mighty Columbia River would provide the necessary amounts of low-temperature water to cool the chain-reacting piles. The Hanford construction camp that housed a population of 51,000 at its peak in 1945 was abandoned after plant construction was completed that same year.

The housing and business development for the Hanford Engineer Works in 1945—now a "ghost" town.

It was apparent from many standpoints that General Groves and his staff and Crawford H. Greenewalt and his Du Pont engineers had chosen well in the matter of a site. But, in addition, it was apparent that they were wasting no time in turning this lonesome expanse into a beehive of special activity. At the Hanford camp, rows of barracks, tents, and trailers stretched in all directions. Tons of materials and construction equipment, ordinarily scarce during the war, were pouring into the area. And even at this early stage I recall eating in the largest mess hall I had ever seen. Certainly I was impressed with the scope of the project, and it was all the more amazing when one considered its purpose—to conduct large-scale alchemy. Here we were going to produce kilogram amounts of a new element that we had been able to isolate in a weighable quantity only a few months before. This first weighable amount was only 2.77 micrograms. As many of you may recall, that first weighing took place in Room 405 Jones Laboratory at the University of Chicago on Sept. 10, 1942. It was carried out on a specially constructed balance using an extremely thin quartz fiber, and the sample was so small that some people still say our experiment consisted of weighing an invisible amount on an invisible balance. Incidentally, last fall we reenacted the first weighing of plutonium at a reunion in Chicago commemorating the twenty-fifth anniversary of the original event.

Looking back on the history of Hanford, we see so much to indicate that in a sense it was a remarkable act of faith and a result of the determination and courage of many men—I wish there were time to mention all their names and give each the individual credit he deserves.

The Hanford project represented the largest scale-up of industrial production ever attempted by man. From the ultramicrochemical experiments we had performed in Chicago to the final Hanford plant production, the amounts of plutonium handled would represent a scale-up by a factor of 10^9. Of course, that is only one phase of the

One of eight Hanford mess halls. Food preparation was on a scale surpassing all but the largest army camps.

large-scale alchemy of plutonium production. If we consider the future production of this element and compare these projected amounts with those involved in the very first tracer studies of the chemistry of plutonium, the scale-up becomes all the more incredible. We are then considering an escalation from picogram amounts (10^{-12} gram) to quantities measured in hundreds of metric tons. Thus the scale-up will amount to about 10^{20}, a number more meaningful to astrophysicists than to chemists!

This scale-up possibility is due to plutonium's unique promise of service to mankind. As used in breeder reactors, it will be the fuel of the future. And it will serve as an energy source in devices ranging from a surgically implanted artificial heart to scientific and life-supporting units in deep space. In fact, there has been some talk that the future value of plutonium may someday make it a logical contender to replace gold as the standard of our monetary system. However, not being an economist, I will not try at this time to predict how we would operate on this new "plutonium standard."

Let us go from the speculative future back to the interesting past and look at the original Hanford Project another way. Fermi's first reactor in Chicago produced less than a single watt of thermal power. The reactors built here at Hanford were designed to generate 200,000 kilowatts. If Fermi's pile had operated continuously for millions of years, it would not have produced the plutonium made in one year in one Hanford reactor. Perhaps even more remarkable at the time was the fact that construction of the reactor plant was begun on June 7, 1943, when just six days earlier, on June 1, 1943, we had decided that the bismuth phosphate process developed by Stanley G. Thompson should be the method for chemical separation of the plutonium that would be made in these reactors. Thus we passed another important twenty-fifth anniversary just last Saturday, another indication that we have chosen an appropriate time for this celebration.

That decision to use the bismuth phosphate process was made in an important all-day meeting in Chicago, and, as some of you may recall, it was a difficult decision to make. Attending the meeting were representatives of the Du Pont Company and the Metallurgical Laboratory, together with the chemists and engineers who had been most closely associated with the development of chemical processes. We summarized all the available information relative to the bismuth phosphate process and the lanthanum fluoride process, which was the other leading contender at the time. Even with all the available data we had on these two processes there remained uncertainties as to the final choice. For example, there were the corrosion problems associated with the lanthanum fluoride process, and there was the possible failure of complete coprecipitation of plutonium with bismuth phosphate at high concentrations of plutonium. In fact, it was not even certain that precipitation processes would necessarily prove superior to methods based on solvent extraction, adsorption, volatility, or other phenomena. But the bismuth phosphate process won out, largely because of my guarantee of at least a 50% yield of chemically extracted plutonium and because of Crawford Greenewalt's preference for a process that would minimize possible early equipment failures even at the expense of possible decreased yields; the potential of corrosion of equipment with the lanthanum fluoride process presented the possibility of complete failure.

In retrospect, it is somewhat amazing that this deliberation took place only a week before actual construction began on the Hanford reactors.

I recall that during my second trip out here, in December 1944, I was asked to sign the numerous specifications for the bismuth phosphate process to be operated in the just-completed first chemical separation plant. This was, amazingly, only 18 months after the historic June 1 decision. Equally amazing was the rate at which the chemical separation plant went into operation. We received the operating standards on Dec. 15, 1944. The first production run of Hanford material in this chemical separation plant began nine days later (Dec. 26, 1944), and, in less than two months (Feb. 2, 1945), the first delivery of plutonium was made from Hanford to the Los Alamos Laboratory.

As I think back to the Hanford–Richland area of 25 years ago and compare it to what we see here today, I am reminded of changes in things other than the plutonium production facilities. For one thing—if I might digress from nuclear matters—there is quite an improvement over the transportation I encountered on my first visit here, in late May and early June 1944. As I recall, John Willard and I had as our host and personal guide on that visit Walter O. Simon. Walter was one of Hanford's leading citizens, having the double honor of being plant manager and the mayor of Hanford. However—and I hope Walter will bear me out on this—one trip he took us on showed that his auto driving did not quite match his managerial talents. In the evening, as he drove us from the site back to the guest house in Richland, he became impatient with the secondary roads in the area (there were no other kind anyway) and decided to take a shortcut straight across the desert. Sure enough, in an area where one might expect to come across a cattle skull or Gabby Hayes, the wheels of Walter's car got trapped in loose sand.

The first Hanford pile. . .

The just-completed first pile area designated 100B. The pile building is the concrete structure in the center, a windowless monolith towering 120 feet above the desert. Behind it lies the water-treatment facilities. The Columbia River and the Wahluke Slope are in the background.

Laying graphite in the pile.

At the front or operating face of the pile, workmen are loading uranium in it.

The just-completed first Chemical Separation Plant. . .

The entire area above the cells was enclosed by a single gallery 60 feet high and running the length of the building. Radiation meant remote control and remote maintenance. Once the plant was operating, the only access to the cells would be by means of a huge bridge crane that traveled the length of the building.

A row of 40 concrete cells, each separated from its neighbor by 6 feet of concrete, most of them about 15 feet square and 20 feet deep, ran the length of the 800-foot-long chemical-separation building.

Stanley G. Thompson (a 1970 photograph) developed the bismuth phosphate process for chemically separating plutonium from uranium. He is shown here with a model that suggests the possibility of superheavy elements. Dr. Thompson points to a representation of eka-platinum-294, for which he has been searching in nature.

The roads in and out of Richland have been considerably improved since the early years, but the scene on either side remains the same, a vast sea of sagebrush and cheat grass.

Fortunately we discovered a water hole nearby, and John and I carried water over to pack down the sand, hoping to create a little better traction. Then we also discovered that Walter was somewhat of an expert at spinning his wheels. He impatiently dug the car in deeper and deeper before we could finish our compaction of the sand. With a great deal of persuasion and some restraint, John and I finally succeeded in slowing Walter down. We managed to get free and make our way back to town—but not before there were a few buzzards circling overhead, or so we imagined. They tell me Walter's driving has improved since then, and so have the roads in and out of Richland.

Richland, of course, has changed in a great many ways since that time. And it is particularly appropriate to take note of its progress during this anniversary year. Richland is a community that grew naturally from the needs of the scientists, engineers, and administrators who created and operated the Hanford Works. It is also an example of community growth based on the diversity that springs from a great scientific and technical enterprise and is nurtured and encouraged by a progressive, energetic citizenry.

Some people might say that today's Richland is representative of "the town that science built." I would go one step further and say it represents the human bonus that results from our Nuclear Age endeavors—a type of spin-off for people which goes with our technological progress. I am pleased to say that the atom has been responsible for some healthy and happy communities, and Richland is certainly among the foremost of them. Of course, no small credit for this is due to your enlightened and enterprising community leaders—your mayor and public officials and your regional organizations, such as the Tri-Cities Nuclear Council—and certainly to your fine Senators Warren Magnuson and Henry Jackson and outstanding Congresswoman Mrs. Catherine May, who have always worked so hard in your behalf.

All their efforts in cooperation with private industry and the federal government have set the pattern for making Richland a city of the future. I believe that the changes fostered here are the results of the kind of thinking and long-range planning that will be essential in our country in the coming years, when rapid economic and social change must take place to meet new national goals.

Returning to the broader aspects of our celebration today, what have all these pioneering years of Hanford and Richland seen—to what overall effort have they contributed? They have contributed far more than the massive production of a new chemical element and the technologies and spin-off associated with it. I believe they have seen the evolution of a new age, the full possibilities of which we are now only beginning to realize. The great potential power from the nuclear furnaces of Hanford is now becoming the fire of the future. It is a type of fire that will bring more people not only physical comforts—through heat, light, power, water, and food—but also a large measure of added knowledge and understanding. The greater significance behind the large-scale alchemy that has taken place at Hanford is that of being able to bring forth changes in the lives of people and the healthy growth of their communities through scientific and technological changes. It also involves the future production of energy on such a grand scale and so cheaply that we may well see an era when such a material as plutonium will radically change our relations to almost all other materials—to our production of food and the use we make of our water, air, minerals, and other natural resources.

The changes that will be wrought by nuclear power are, of course, dependent on farsighted economic and environmental thinking, and in this respect we in the Nuclear Age have a decided advantage over previous generations. This is an age when we have the ability, the tools, and the wisdom to look ahead and plan wisely for the future.

Prior to this time our use of energy resources has been rather haphazard. We have used our natural fuels—wood, coal, gas, and oil—as we have discovered them and needed them, without much long-range thinking and planning. Fortunately our Nuclear Age is one of foresight as well as power, and therefore we are thinking in terms of conserving our energy resources, of using them as efficiently and economically as possible, and of their relation to all our other resources and our total environment. This new level of thinking is perhaps the most significant phenomenon that has taken place paralleling our current technological revolution. As a result, in the nuclear field we are not content to sit back and watch the growing use of our present range of nuclear power stations, no matter how safe, reliable, and economically competitive they may be today and for the coming decade or so. We are already very hard at work developing the more efficient converter reactors and the breeder reactors of the future which will allow us to make far more efficient and economic use of our natural nuclear resources.

On Feb. 28, 1950, a meeting with Hanford area officials at the time of Glenn T. Seaborg's visit to Hanford as a member of the General Advisory Committee. (Left to right) A. B. Greninger, Manager of the Hanford Technical Division; George R. Prout, Hanford General Manager; Seaborg; Donald G. Sturgis, Director of the Hanford Production Division; and Fred C. Schlemmer, AEC Manager.

The Redox plant takes shape at Hanford in December 1950. The long "canyon" of concrete cells would contain chemical equipment for recovering plutonium and uranium from slugs irradiated in the Hanford production reactors. (Redox replaced the bismuth phosphate plant for plutonium separation.)

Richland, of course, will be highly involved in the development of an important kind of breeder reactor, the fast breeder, as the Fast Flux Test Facility will be located here. This facility, which will play a vital role in the development of the fast breeder through the testing of its fuels and materials, is an example of how nuclear development can bring additional growth to a community.

Perhaps 25 years from now we will be able to gather here to look back over half a century of progress of the Nuclear Age. By then Richland, together with the Tri-Cities Area, will probably be a large metropolis thriving on its growing science-based industries. Perhaps Hanford will be its Nuplex, able to preserve the surrounding vast and majestic area close to the way nature created it. We will be able to reminisce about the beginning of the Nuclear Age while we see all about us many of the wonders that it has brought and continues to unfold. In the meantime, I think that all the citizens of Richland and those who helped to make history at Hanford can be justifiably proud today. ■

URANIUM-233

A Generation After Its Discovery A Future of Promise

On Oct. 8, 1968, Glenn T. Seaborg brought the Molten Salt Reactor Experiment to power on a fuel charge of uranium-233, making it the world's first reactor to operate on uranium-233 fuel.

First Uranium-233 Loading of the Molten Salt Reactor Experiment, Oak Ridge, Tennessee, Oct. 8, 1968

■ Today we are recognizing another eventful step in nuclear progress made here at Oak Ridge. The first loading of the Molten Salt Reactor Experiment with uranium-233 fuel is an important development in evaluating the potential of molten-salt reactor systems for the commercial production of electricity.

No matter how many times I come to Oak Ridge I am always impressed by this vast facility for scientific discovery and technological development. And, when I reflect on the history of uranium-233 and think about how important it is going to become in the future, Oak Ridge, along with the rest of the great public and private nuclear establishments, seems even more remarkable. What we are observing today is the union of uranium-233 with advanced reactor technology—the merger of an isotope discovered more than a quarter of a century ago as the result of what today would be a real shoestring program with a concept and experiment of contemporary technology that promises to open new vistas for producing and using energy.

The primary significance of uranium-233 is that this man-made element opens up the vast amount of energy stored in thorium-232, which is so abundant in nature. Uranium-233 and plutonium-239 (which opens up the abundant but nonfissionable isotope uranium-238), together with the relatively more scarce but fissionable isotope uranium-235, make the nucleus of the atom a virtually unlimited source of energy. Just as it is possible to visualize the world running out of fossil fuels, it is also possible to visualize running out of economically recoverable uranium-235. But it is not easy to conceive of exhausting nature's supply of uranium-238 or thorium-232, each of which is about 100 times as abundant as uranium-235. In the United States alone, the value of energy that could be derived from our reserves of uranium-238 and thorium-232 through the plutonium-239 and uranium-233 fuel cycles, measured in terms of what it would cost to produce an equivalent amount of energy from fossil fuels, is probably something on the order of 50 quadrillion dollars for each of those two fuels.

The key to making the most effective use of nature's supply of nuclear fuels is, of course, the breeder reactor. Breeders will make it economical to use relatively high-cost ores as sources of fuel and will, therefore, greatly extend the world's supply of economically usable uranium-238 and thorium-232. The AEC's program to develop breeder reactors now gives the highest priority to the liquid-metal-cooled fast breeder reactor, but we are also investigating other concepts that appear to have an important potential for breeding. One of these is the molten-salt breeder reactor, and the Molten Salt Reactor Experiment is a part of our molten-salt reactor program. The molten-salt reactor, with its thermal-neutron spectrum, fluid-fuel system, and thorium–uranium-233 fuel cycle, could turn out to be a strong system for parallel development or an excellent backup to the liquid-metal fast breeder reactor, with its fast-neutron spectrum, solid-fuel system, and uranium–plutonium fuel cycle.

In loading uranium-233 in the Molten Salt Reactor Experiment, we are taking an important step toward developing many of the materials and design features of breeders for actual use in an operating reactor. At the present stage of development, the molten-salt breeder concept merits investigation for a number of important reasons. Its fuel-inventory costs would be low, and there would be essentially no cost for fuel fabrication. So far the Molten Salt Reactor Experiment has operated successfully and has earned a reputation for reliability. Although too much development work remains to be done on breeder reactors for us to make confident predictions, some people believe that the molten-salt thermal concept will provide the quickest and the least expensive route to competitive commercial power from breeder reactors. In any event, I think that someday the world will have commercial power reactors of both the uranium–plutonium and the thorium–uranium types.

Regardless of which breeder concept or concepts eventually

The Molten Salt Reactor Experiment in its 24-foot containment cell, which is covered during operation.

In an agro-industrial complex, as envisioned in this artist's conception, large nuclear reactors would provide desalted water for agriculture and generate electricity for industry.

become the basis for future commercial nuclear power stations, the advent of the breeder will have far-reaching effects on the production, cost, and utilization of energy. With low-cost nuclear power available in large amounts, as Alvin Weinberg has pointed out, a new age of energy-intensive industrial technology could be opened up. Some of the concepts that are being studied here at Oak Ridge, such as the agro-industrial complex investigated by E. A. Mason, R. P. Hammond, and their many able associates, could be brought closer to reality.

Back in 1941 and 1942, during the work leading to the discovery of plutonium and uranium-233, it would have been difficult to visualize the progress that has occurred up to now. At that time not many scientists would have believed that the nuclear establishments of both the federal government and private enterprise would have grown to be as large and diversified as they are today, that nuclear power plants would be competitive with fossil-fueled power plants in the generation of electricity, that nuclear energy would be a practical and economic reality in so many other different ways, or that in only a generation after 1942 we would be within 15 years or so of producing commercial electricity in breeder reactors.

I am tremendously gratified at seeing uranium-233, discovered 27 years ago, being more closely linked with the great energy technology of the future. We are now on the threshold of making tremendous advances in the amount of energy that can be obtained economically from nature. This first loading of the Molten Salt Reactor Experiment with uranium-233 may lead to some highly significant developments in the history of nuclear energy technology. Those of you who have worked on this molten-salt reactor project have our praise and our gratitude for a job well done. ■

LOS ALAMOS

25 Years in the Service of Science and the Nation

Twenty-fifth Anniversary Celebration of the Los Alamos Scientific Laboratory, Los Alamos, New Mexico, Feb. 15, 1968

■ Within the past few months we have been reminded by several anniversaries that the Nuclear Age is a quarter of a century old. But, no matter from what single event we date the birth of the Nuclear Age, we know the Los Alamos Scientific Laboratory played a leading role in that birth and, perhaps more significant today, continues to advance the Nuclear Age as one of the nation's leading nuclear laboratories. It is, therefore, in the spirit of both celebration and expectation—looking back at a productive past and ahead to a promising future—that I would like to speak this evening.

Standing here in this large auditorium tonight, I find it difficult to believe that 25 years ago this site contained nothing but a few barns and stables of the Los Alamos ranch school. To the east we can still see much of the dense piñon forest that covered the slope up to the ranch school at the eastern end of the mesa. The transformation of this pastoral setting into a modern, bustling community began in the summer of 1942 when Robert Oppenheimer, in the company of Gen. Leslie R. Groves, returned to the school he had often visited on pack trips from his summer home across the Rio Grande Valley.

After visiting Jemez Springs and other possible sites in the region, the two men decided that the ranch school would be the best place to establish a small laboratory for designing and building the world's first nuclear weapon. The War Department took over the school in December 1942, and by January 1943 General Groves had formally established the Los Alamos Laboratory, known then as Project Y. By the end of February, the school personnel had left the premises, and another saga of American pioneer history began as scientists this

Twenty-five years ago
this site contained nothing
but a few barns and stables
of the Los Alamos ranch school. . .

To the east, the dense pinon forest that covered the slope up to the ranch school.

Ranch-school boys canoeing on Ashley Pond before the land was taken over for the laboratory.

A ranch-school graduation ceremony.

Scene on an early homestead in the area.

time rather than ranchers sought, with Army assistance, to create a new community in the vast reaches of the West.

Many of you, I am sure, remember those exciting if uncomfortable first months on the Hill—the dirt, the construction, the trailers and plywood buildings, the lack of water and telephone service, the barracks, and the high security fences. Inside, despite the chaos and inconvenience, Oppenheimer and his associates were rapidly building one of the most illustrious scientific laboratories in the world or, as General Groves is said to have jokingly described it, "the greatest collection of crackpots the world has ever seen." The Los Alamos

The dirt, the trailers and plywood buildings, the barracks, and the high security fences. . .

roster in those days read like a scientific who's who: Niels Bohr, Oppenheimer, Hans Bethe, Enrico Fermi, Robert Bacher, Edward Teller, George Kistiakowsky, and a score of talented young men and women who were destined to leave a lasting mark in the history of American science.

To a large extent the greatness of Los Alamos lay in Robert Oppenheimer. An exceptional theoretical physicist and an extraordinary teacher, he provided the spirit and inspiration needed for the incredible job the laboratory faced. As a young chemist at Berkeley in the 1930s, I spent many hours with Oppie discussing scientific problems I then thought important. Whatever he thought of my questions, he always accorded me a full measure of patience and understanding, and he earned in my estimation a unique kind of admiration and respect. I saw him only occasionally during the war when he visited the Metallurgical Laboratory in Chicago, but I know from what many have said that the same qualities that inspired us at Berkeley provided the motive force at Los Alamos.

Enrico Fermi supervises extraction of L. D. P. King's car from the snow by means of a horse team.

Although I never visited Los Alamos during World War II, I naturally had a keen interest in the research going on here, particularly in the chemistry and metallurgy of plutonium. Joseph Kennedy and Arthur Wahl, who had worked with me at Berkeley in first identifying the element plutonium, carried much of the plutonium effort at Los Alamos. Because of Kennedy's extreme youth at the time, authorities in Washington questioned Oppenheimer's wisdom in appointing him head of LASL's Chemistry and Metallurgy Division, which was responsible for final purification of plutonium and uranium-235, but Kennedy's abilities were too obvious to be denied.

I especially remember the work done at Los Alamos in the summer of 1943 in determining whether plutonium-239 emitted neutrons in the process of fissioning. If it had not, it would have had no value for weapons. Robert Wilson, now director of the National Accelerator Laboratory at Weston, Illinois, borrowed almost all the plutonium we had in Chicago at that time (then about 200 micrograms) for the all-important plutonium fission experiments at Los Alamos. I recall very well that he completed his experiment

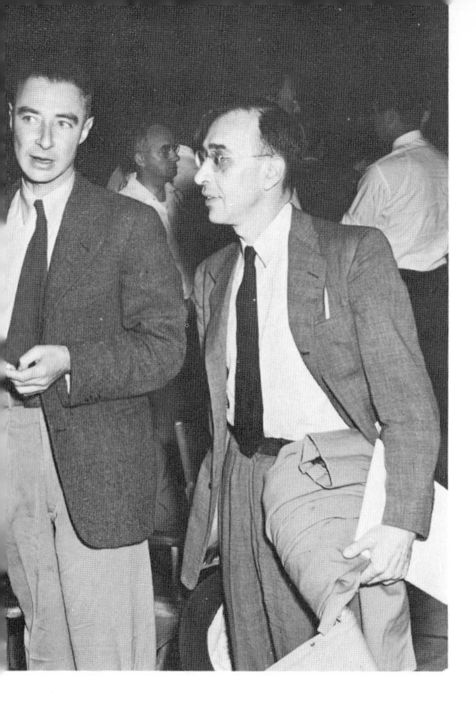

J. Robert Oppenheimer and Greg Breit, at the conclusion of a LASL technical symposium.

Inside, despite the chaos and inconvenience, Oppenheimer and his associates were rapidly building one of the most illustrious scientific laboratories in the world. . .

The original Tech Area.

while I was in Santa Fe on a short vacation. Before dawn one morning, with a rifle in his car for security protection, Wilson brought the plutonium sample to Santa Fe and met Mrs. Seaborg and me in a restaurant. I then carried the sample in my suitcase on my train ride the rest of the way to Chicago without benefit of firearms. It may never be possible again to use such informal methods of transfer, but they served their purpose in the exigencies of war.

Important to the successful development of the nuclear weapons was the work in the fast-neutron physics connected with uranium-235 and plutonium-239. The contribution of John Manley and his fast-fission team from Chicago and the work of Alvin C. Graves and James Coon were significant during this time.

During most of those legendary 28 months of Los Alamos's history of World War II, plutonium seemed to be the hinge of fate. The splendid work done by Navy Captain William S. Parsons, my former colleague Edwin McMillan, and the people working with them seemed to guarantee the success of the guntype weapon using uranium-235. But the plutonium approach always seemed to be in trouble. Although we were confident after Wilson's experiments that plutonium would provide enough fission neutrons to sustain the chain reaction, we still were not sure that we could produce plutonium pure enough to use in a guntype weapon. With encouragement from John von Neumann and Edward Teller in the fall of 1943, Parsons and his associates began to supplement the earlier work of Seth Neddermeyer on the implosion method of assembly.

One of the practical, if primitive, ways in which radioactive materials were safely handled during the wartime period in which the first atomic bombs were being developed.

The difficulties of developing the implosion method were not a matter of great concern until July 1944, when the first samples of plutonium arrived in Los Alamos from the X-10 Reactor at Oak Ridge. Unlike the small samples we had obtained from cyclotron runs, the reactor-produced plutonium contained a significant amount of plutonium-240, whose high spontaneous fission rate, discovered by Emilio Segrè and co-workers, made it impractical for use in a guntype weapon. Joe Kennedy, Cyril Smith, and their associates were by this time confident that they could remove most of the impurities in plutonium, but they could not hope to develop a separation process for extracting the plutonium-240 isotope. If plutonium was to have any use during the war, Los Alamos would have to solve the implosion puzzle. Taking this disappointment in stride, Oppenheimer reorganized his forces to concentrate on the implosion system. Bacher and Kistiakowsky put full time on the effort; Bethe helped on theoretical problems; and Samuel K. Allison coordinated the work of all groups working on implosion. A year later, almost to the day, Los Alamos was successful. On July 16, 1945, the world's first nuclear detonation, produced by an implosion device using plutonium, occurred at Alamogordo.

Within a few weeks the triumphant scientific success of Trinity was tempered by the awesome news from Hiroshima and Nagasaki. The feverish work of the past few years ended almost overnight. As the world began to comprehend the destructive power of the atom, some prominent Americans voiced the opinion that war was no longer conceivable. If there was to be a future at all for any nuclear weapons laboratory, it was not likely to be for one located on a

The icehouse, a stone building long used before the war for storage of ice from Ashley Pond. Nuclear components of the first atomic bombs were checked and assembled here.

On July 16, 1945,
the worlds first nuclear detonation
occurred at Alamogordo. . .

Active material for the Trinity device is moved from the sedan that brought it to McDonald ranch.

McDonald ranch, used for final assembly of the Trinity device.

The first atomic device waits in its steel shelter at the top of the tower, ready for countdown.

The main control point for the Trinity test. Robert Oppenheimer and General Farrell were among those who watched the explosion from this bunker.

Trinity shot, July 16, 1945.

Within a few weeks
the scientific success of Trinity
was tempered by the awesome news
from Hiroshima and Nagasaki. . .

Above: The "Fat Man," a plutonium bomb similar to those detonated over Trinity and over Nagasaki. Below: The "Little Boy," a uranium bomb similar to that detonated over Hiroshima.

remote mesa in New Mexico after the veil of wartime secrecy had been torn away. Before the summer of 1945 ended, the spectacular constellation of scientific talent at Los Alamos began to disintegrate. Former professors were eager to return to their universities, and their wives were just as anxious to rediscover the joys of central heating, modern kitchens, and urban living. Oppenheimer's departure seemed to mark the end of Los Alamos's brief moment in history.

General Groves did his best in 1946 to hold Los Alamos together. The Army upgraded the unreliable water supply system, started building some permanent homes in the western area, and made plans for a shopping center near the lodge. The continuing existence of Los Alamos was far from certain even by the end of 1946, however, when the new Atomic Energy Commission took over the project from the Army. Los Alamos was still a military reservation, and there were no assurances that the new civilian Commission would see a need for the laboratory on the Hill.

I recall that when I joined the Commission's first General Advisory Committee (GAC) as the most junior member of an otherwise illustrious group Los Alamos was one of the first subjects of discussion. Some of my colleagues maintained that it would never be possible to make Los Alamos attractive for competent scientists.

In the spring of 1946, Director Norris Bradbury and his first tech board met with Gen. L. R. Groves to discuss building permanent homes. Left to right: standing, Col. Herbert Gee, later post commander; Col. L. E. Seaman, post commander and assistant director for administration; Darol K. Froman; Max Roy; John Manley; R. D. Richtmyer. Seated, Bradbury, Groves and Eric Jette.

It was too remote from civilization. The wartime buildings were already falling to pieces, and building permanent structures in such an isolated spot would be too costly. Furthermore, most of the "big name" scientists had left Los Alamos with Oppenheimer. Those remaining might be competent young men with more than average ability, but they could hardly be compared to the giants of the war years. To be specific, some of the GAC members questioned the capabilities of the young Navy Commander who had succeeded Oppenheimer as director. Norris Bradbury was an excellent physicist and had done an outstanding job on the Trinity test, but could he fill Oppenheimer's shoes? In early 1947 at least a substantial minority of the GAC believed that neither Los Alamos nor Norris Bradbury would long be on the atomic energy scene.

That this unhappy prognostication did not come true is, I think, more a tribute to Bradbury and his Los Alamos team than a reflection on the prophesying ability of the GAC. Early in 1947 Oppenheimer himself visited Los Alamos with some members of the GAC. They reported back to the rest of us that there were indeed

Norris Bradbury, the young Navy Commander who succeeded Oppenheimer as Director of the Los Alamos Scientific Laboratory.

problems at Los Alamos, but Bradbury and his staff had demonstrated a determination to push ahead with weapon development despite these difficulties. At that time you may remember there was no such thing as a ready stockpile of nuclear weapons. There were some components of the two wartime types of weapons, but they were custom-made and not adaptable to production-line methods. I remember that during most of 1947 Los Alamos had to devote most of its energies to producing components until new facilities could be constructed elsewhere.

Despite these necessary diversions, Los Alamos never lost sight of its primary mission as a research and development center. Before the end of 1947, Sandia was able to take over many of the assembly and testing jobs on weapons, and the laboratory turned to developing new weapon models and preparing for the Sandstone tests in the Pacific. The stunning success of those tests in the spring of 1948 made it perfectly clear that Bradbury, assisted by Darol Froman, Max Roy, John Manley, Robert Richtmeyer, Raemer Schrieber, Eric Jette, Jerry Kellog, Al Graves, Marshall Holloway, and many others, had firmly reestablished the excellence of Los Alamos as a research center. Sandstone opened a new era for weapons development, and Los Alamos moved ahead quickly on several fronts to exploit various possibilities for building more efficient and more reliable weapons.

At the same time, Bradbury and his staff were planning other kinds of research with less direct application to immediate weapons requirements. Following pioneering work started in 1946 by Philip Morrison and Louis Slotin and continued by David and Jane Hall, the fast-neutron reactor Clementine went critical in 1947. A report to Washington in September 1948 proposed research with Clementine, basic studies of plutonium and tritium, construction of an electronic computer, continuing theoretical studies on various approaches to a thermonuclear weapon, and further investigation of advanced weapon design. Basic research in nuclear physics, chemistry, and biology would complete the transformation of Los Alamos from a task force with a narrow mission to an applied physics laboratory. A special Commission consultant, after visiting Los Alamos in the spring of 1949, reported that it was the finest government laboratory in the nation, a tribute to Bradbury and those who worked with him.

The revitalization of Los Alamos seemed to come none too soon. In September 1949 the United States first detected a nuclear detonation in the Soviet Union. Early in 1950 came the revelation of Klaus Fuchs's espionage activities, and a few months later the

Korean War began. In February 1950 President Truman ordered the Commission to accelerate research and development on a thermonuclear weapon. Naturally the burden fell largely on Los Alamos. Carson Mark and the theoretical physics division, building on earlier theoretical studies by John von Neumann, Stanley Ulam, and Edward Teller, turned most of their attention to this problem. In little more than a year, Teller and Ulam proposed what seemed to be a practical design. Froman and Holloway organized a group in the laboratory to build the test device detonated at Eniwetok in November 1952. Less than two years of concerted effort had produced a new weapon of unprecedented power.

During these same years the laboratory as a whole was devoting most of its efforts to developing a family of fission weapons for a variety of uses, all the way from large strategic weapons to artillery shells and underwater bombs. In the early 1950s the Nevada Test Site, created largely by Los Alamos personnel, first became a key installation in weapon development. The increasing tempo of weapons tests in those years is a rough indicator of the rapidly growing capability of Los Alamos in designing and testing new weapon models.

For nearly 20 years Ground Zero at Trinity site was identified simply by an old board sign.

A lone sign on the road to Journada del Muerto.

Above: Clementine, the world's first plutonium reactor and first fast-neutron reactor. Below: The Water Boiler, the world's first homogeneous reactor and the first reactor to use enriched uranium. Both of these reactors were completed in the first few years of LASL's first decade.

As a full-scale development laboratory, Los Alamos inevitably required the tools for fundamental research. Just as weapons research requirements had led to the construction of Lopo, the world's first homogeneous reactor, and Clementine, the world's first fast-neutron reactor, so the laboratory in the 1950s continued to have a leading role in reactor development. In 1956 the Omega West Reactor replaced Clementine. Like its predecessors, OWR was primarily a neutron source, but the growing interest in nuclear power also had its effect at Los Alamos. Two experimental power reactors, the Los Alamos Plutonium Reactor Experiments (LAPRE) I and II, were extensions of the laboratory's interest in homogeneous reactors, as were the two Los Alamos Molten Plutonium Reactor Experiment (LAMPRE) reactors of the following decade.

Reactors were not the only product of weapons development activities in the 1950s. The intensive effort to create the thermo-

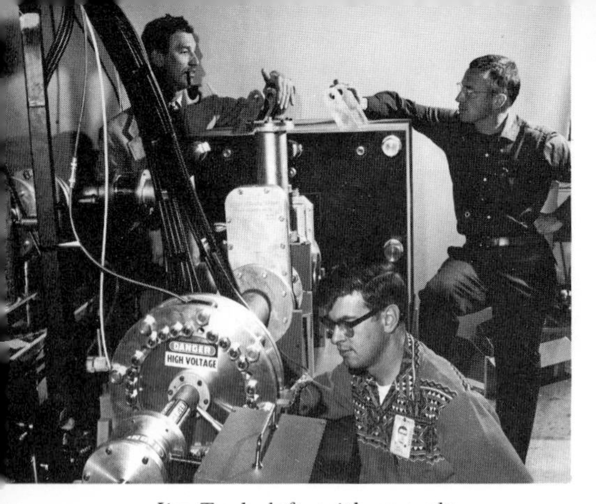

Jim Tuck, left, with an early "picket fence" fusion experiment.

President John F. Kennedy operat "mechanical hands" at the Nuclea Rocket Development Station at Ne Test Site, December 8, 1962. To hi right Glenn T. Seaborg, Space Nuc Propulsion Office Manager Harol Finger, and U.S. Senator Alan Bib (Nevada).

nuclear weapon naturally stirred an interest in controlling the same reaction as a source of power. Just a few weeks before the famous Mike shot at Eniwetok in the fall of 1952, James Tuck first attempted to operate a device for containing a plasma of hydrogen ions. Stressing the experimental nature of the device, Tuck called it the "Perhapsatron." Although it did not work, it did lead to further research on the "pinch effect." By the time of the Geneva conference in 1958, Tuck and his associates had already explored the linear pinch and magnetic mirrors and had built Scylla, which approached thermonuclear temperatures in dramatic fashion in the exhibit hall in Geneva. Similar spin-offs from weapon work occurred during the 1950s in the chemistry and metallurgy division, which completed some pioneering studies of the properties of elements at very low temperatures. Many other examples of outstanding research in physics, chemistry, metallurgy, and biology could be cited, but they hardly seem necessary to demonstrate the breadth and depth of basic research at Los Alamos during those years.

As you know, LASL has been an outstanding contributor to the attainment of major milestones in the nuclear rocket program—the Rover program, including the KIWI effort, which resulted in the first successful full power test; the Phoebus series of higher power tests; and fuels and materials technology activities in support of a nuclear rocket effort. Such an effort is a vital part of our long-range space plans because of the essential role nuclear rockets must play in logistic supply to support extensive manned operations on the moon and eventually to carry out manned missions to the planets. In addition to its work directly on the nuclear rocket reactor, the laboratory has demonstrated outstanding capability in creating major handling and testing facilities that in and of themselves were significant research and development efforts.

Although LASL will remain a major nuclear weapons laboratory for some years to come, I believe that it will also play an increasingly significant role in advancing many areas of our scientific knowledge. I see Los Alamos making great strides in the years ahead toward becoming a "center of excellence" for the entire Rocky Mountain region. Its area of academic research and training can be expected to grow to strengthen both LASL and the academic institutions of this region. In this context I can see Associated Western Universities, with its 17 member universities, and the Los Alamos Scientific Laboratory working in close harmony to advance education in the Rocky Mountain area to the point where it will be someday on a par with the very best the nation has to offer in higher education and graduate training. We have entered an era when all regions of the country must have their centers of excellence, and the development and growth of such centers will have a pervasive influence on the quality of education in each region. National laboratories and national scientific facilities such as Los Alamos can and should expand their cooperative efforts with the educational community. As we find more and more ways to use scientific and technological means and achieve economic and social progress, and as our institutions of higher education become increasingly involved in guiding our society, the partnership between the universities and such government laboratories as Los Alamos should take on a greater significance in fostering the nation's well-being. The decades ahead will see a new and more meaningful era of progress if we can combine and focus our tremendous resources of knowledge and human talent toward common goals.

I have tried today to summarize some of the history and accomplishments of LASL—the accomplishments are far too numerous to do full justice to in so short a time and the history, many of you have lived through quite closely. I have also tried to project something of the future of this great laboratory. On this twenty-fifth anniversary the Los Alamos Scientific Laboratory can celebrate the past with pride and look to the future with hope. You have earned the right to do both. My congratulations and best wishes to every one of you. ■

LIVERMORE

15 Years of Achievement

View of the Livermore Laboratory from the top of the present Cyclotron Building, July 1952. The first group of Berkeley personnel had just begun to prepare the wooden barrackslike buildings at this former Naval Air Base for occupancy.

Fifteenth Anniversary and Dedication of the Radiochemistry Building, Lawrence Radiation Laboratory, Livermore, California, operated by the University of California, Sept. 9, 1967.

■ My reasons for accepting enthusiastically Dr. May's invitation to speak on this occasion are several. First of all, I naturally have a tender spot in my heart for radiochemistry, a field of great importance to my research in nuclear chemistry—and what scientist can resist a busman's holiday! Second, it is always good to see old friends and former colleagues and students, some of whom, I am proud to say, I helped to recruit for this laboratory when it was formed. Finally, this occasion marks the fifteenth anniversary of the Lawrence Radiation Laboratory, Livermore, and it is a pleasure to extend, on behalf of the entire Atomic Energy Commission, congratulations on your extensive and outstanding contributions to the national security and to the expansion of the peaceful potentials of nuclear energy.

At this time it is also a privilege and a pleasure for me to bring

On this 15th anniversary of the Lawrence Radiation Laboratory at Livermore. . .

View of the Livermore Laboratory from the same place, July 1969.

At the dedication, inside the Radiochemistry Building (left to right), Michael M. May, Congressman Craig Hosmer, Gerald Tape, Congressman John Anderson, and Glenn T. Seaborg.

you the following message from the President of the United States:

On this fifteenth anniversary of the Lawrence Radiation Laboratory at Livermore, it is my pleasure to extend, through Chairman Seaborg, my congratulations to the Director and the entire staff of the laboratory.

For 15 years the Lawrence Radiation Laboratory at Livermore has been making an outstanding contribution to our national security and defense. On this occasion I want you to know that your important work, carried on in strict secrecy and unheralded by the American public, is recognized and appreciated by your President and your nation.

Your work has been essential to maintaining world peace as well as to advancing man's knowledge about the atom and its ever-growing potential for progress as well as security. On this anniversary I hope you will all take pride in what has been accomplished by the Lawrence Radiation Laboratory at Livermore and will continue your efforts, which result in the many contributions this laboratory makes to the nation.

That concludes the message from President Lyndon B. Johnson.

The new Radiochemistry Building we are dedicating today and the work that will be done in it derive from a distinguished heritage.

The lineage of radiochemistry—and, more broadly, that of nuclear science and development—can be traced to the discovery of natural radioactivity by Becquerel and the early contributions of the Curies. This work, starting in 1896, opened the door to the atomic nucleus. The line of descent leading to modern radiochemistry started especially with Madame Curie, who patiently applied fractional crystallization techniques to the separation and identification of minute sources of radioactivity in large quantities of pitchblendes. In particular, her brilliant discovery, isolation, and clear demonstration of the chemical properties of radium constituted a model to which those of us who followed look with admiration and awe.

Otto Hahn, whose illustrious work spanned half a century, is widely regarded as the father of modern radiochemistry. It was his bone-deep knowledge of chemical theory and his mastery of radiochemistry techniques that led Hahn and his coworker Strassmann to the historic discovery of nuclear fission. The methods Hahn and others had developed in work on natural radioactivity were of immense value when artificial radioactivity was discovered in 1934. Among those prominent in extending the techniques in the field were members of the Berkeley school of nuclear chemistry which

The early contributions of the Curies. . .

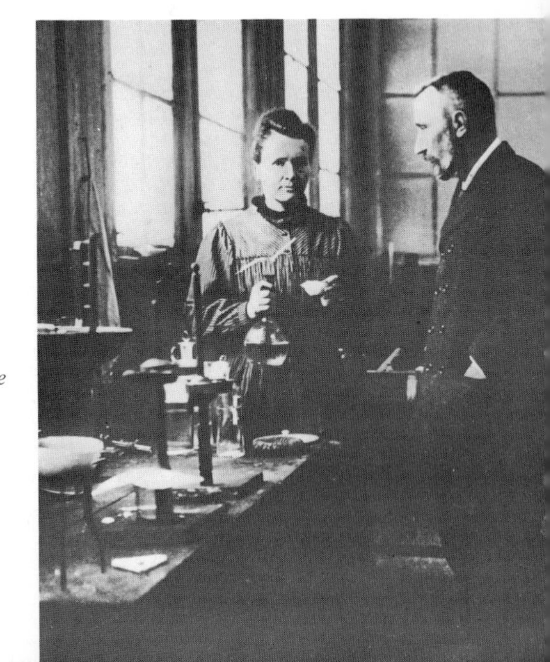

Marie and Pierre Curie in their laboratory.

The Berkeley school of nuclear chemistry developed largely because of the cyclotrons of the late Ernest O. Lawrence. . .

In 1931 Ernest O. Lawrence (right), inventor of the cyclotron, and M. Stanley Livingston, his student, inspect a 74-ton magnet. The magnet, a "white elephant" built for radio transmission but never used, was given to Lawrence by the Federal Telegraph Company and was installed in Lawrence's laboratory at the University of California, Berkeley, in the fall of 1931. Using this magnet, Lawrence, Livingston, and their colleagues constructed the first major cyclotron. Initially a 27-inch accelerating chamber was installed in the magnet, and in 1937 a 37-inch chamber was installed.

The 16-million electron volt deuteron beam of the Lawrence 60-inch cyclotron at the University of California, Berkeley, shortly after the cyclotron began operating, June 1939. By use of the beam, the new element neptunium was synthesized and discovered in the spring of 1940 and the element plutonium in February 1941.

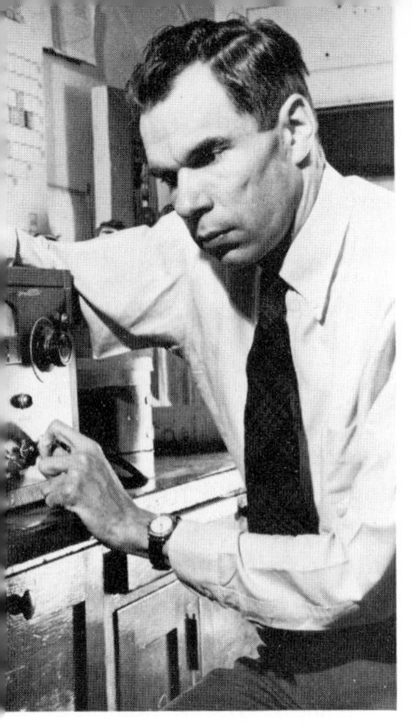

Glenn T. Seaborg counting a sample of radioactive material in an old-fashioned Geiger counter at about the time he and a group of colleagues discovered element 94, plutonium, in 1941.

Glenn T. Seaborg and the elution apparatus used to separate newly created transuranium elements, November 1951.

The tools of radiochemistry were essential in the extension of the periodic table beyond uranium. The discovery of plutonium was an immensely important practical product of this work. . .

developed largely because of the presence of the cyclotrons of the late Ernest O. Lawrence, for whom both the Livermore and Berkeley laboratories are, of course, named.

The tools of radiochemistry, fashioned initially by physicists and chemists and ultimately with strong influence from engineers, were essential in the extension of the periodic table beyond uranium by Berkeley scientists. The discovery of plutonium was an immensely important practical product of this work. The techniques developed in basic research were applied and improved in the emergency of World War II for the separation of plutonium from the fission products generated in the great production reactors.

Subsequently radiochemistry, growing constantly in sophistication, has been indispensable to science and society. The expanding arsenal of tools has helped the nuclear scientist to elucidate nuclear structure in fine detail and to work out with precision the complicated decay schemes of the heavy elements, the fission products, and other interesting nuclei. These techniques have also been indispensable to the development of nuclear power, to the production and widespread use of radioisotopes, and to the elaboration of safety measures that have made the nuclear industry one of the safest we know. Finally, the accumulated knowledge and techniques have been applied and extended here both to assist in the improvement of the nuclear weapons that are important to our national security and to help work out a safe technology for future use in the Plowshare program.

In capability for the application of radiochemistry techniques, this new laboratory probably is unsurpassed anywhere. Indeed, it is somewhat amusing to compare one of my earliest experiments in the use of radiochemistry with the powerful methods available in this building. In the middle 1930s in Berkeley, the main nuclear facility was the Old Radiation Laboratory, a rather ramshackle frame building that housed the 37-inch cyclotron. I was a graduate student in chemistry with a keen interest in the exciting work at the cyclotron. One day Jack Livingood, a physicist with whom I had become acquainted, confronted me with a target he had just bombarded. He literally handed me the hot target and asked me to do some chemistry on it to identify the radioisotopes that had been produced. Naturally I jumped at the chance. The facility he offered in Le Conte Hall was hardly luxurious. My best recollection is that it was the custodian's closet and that the resources consisted of tap water, a sink, and a small work bench. With some essential materials bootlegged from the Department of Chemistry, I performed the chemical separation to Livingood's satisfaction. For several years afterward we worked closely together in searching for new radioisotopes.

This new laboratory and its equipment represent a world far removed from the horse-and-buggy radiochemistry I first encountered some three decades ago. It is symbolic of the rapid movement of science and technology in the intervening thirty years and also of the heavy responsibility borne by those who work here. The sophisticated equipment includes automated alpha-, beta-, and gamma-ray counters and spectrometers designed to accumulate data in a form directly usable for analysis in the extensive computer facility of the laboratory. There are a computer-controlled gamma-ray spectrometer, a milligram-scale isotope separator, and mass spectrometers. The latest handling equipment and built-in shielding will facilitate work and provide safety for personnel.

In the completion of this facility, teamwork of private industry, the Livermore laboratory, and the local staff of the Atomic Energy Commission has been important. The laboratory team was headed by Dr. Walter Nervik, who played a major role in the planning and design, especially in developing the technical requirements and seeing to their implementation.

The rapid and efficient analysis by the Radiochemistry Division of the spectrum of isotopes produced by nuclear devices helps tell the physicists, mathematicians, and engineers how well a device has performed and provides clues to improvements. Thus Livermore radiochemists have shared significantly in the brilliant achievements of the laboratory in supporting the national security and in the developing technology of Plowshare explosive devices.

In addition, these analyses are of immense value in the work being done by the Bio-Medical Division. A knowledge of the spectrum of radioisotopes produced by an explosion under given conditions helps the biomedical scientists predict what biologically important isotopes will be produced, in what quantities, and their likely distribution in the environment.

It should be emphasized, too, that the talented scientists in the Radiochemistry Division carry on a continuing program of basic research with the fine tools in this facility. The areas of interest include nuclear reactions, nuclear structure, radioisotopes, inorganic chemistry in solution and at high temperatures. The resulting contributions have been extensive, but I should like to mention just two examples with which I am especially familiar since they are in my own field of the heavy elements.

One of the phases of the Plowshare program involves the exploration of methods for producing very heavy elements in

A hemispherical cavity about 75 feet high and 134 to 196 feet across remained from the GNOME underground nuclear explosion. Note man standing on rubble at right center.

underground nuclear explosions. In 1964 the Par device, designed for this purpose, was exploded in Nevada. It represented a breakthrough in that this and similar work at Los Alamos marked the first real success in making some of the heavier isotopes by this method in quantities sufficient for real research. Significant results of studies of the debris here at Livermore were the important observation and study of relatively long-lived fermium-257 (element 100), then the heaviest isotope ever produced by man. I am gratified that two of the scientists in the group performing this work obtained their Ph.D.'s under my direction at Berkeley and were among the early recruits in radiochemistry at Livermore—Richard W. Hoff and E. Kenneth Hulet.

These scientists and their colleagues have followed this line of investigation, in collaboration with scientists at the Berkeley Laboratory, with a new achievement. Recently they were recipients, as were scientists in other laboratories, of a portion of the first sizeable quantity of einsteinium, element 99, produced in the High

Flux Isotope Reactor at Oak Ridge. This reactor has just recently commenced the production of relatively large quantities of the heavier transuranium elements for study by scientists.

The Livermore scientists prepared a 3-microgram target of einsteinium, and just a few weeks ago, Albert Ghiorso, who is in charge of the Heavy Ion Linear Accelerator (HILAC) in the Lawrence Radiation Laboratory, Berkeley, used this accelerator to bombard the einsteinium with helium ions. These scientists have given me the privilege of announcing here for the first time that they have observed, as a product of the bombardment, mendelevium-258 (element 101). This is the heaviest isotope definitely observed by man. The isotopes of elements 102 and 103, discovered at the HILAC, were lighter; that is, they had smaller mass numbers. Mendelevium-258 has a special importance because it has a half-life of about two months, which is unusually long for an element in this region of the periodic table. Because of its long half-life, mendelevium-258 can be made in sufficient quantities to carry on substantial tracer chemical studies. The scientists responsible for this work are Kenneth Hulet, R. W. Lougheed, J. E. Evans, J. D. Brady, R. E. Stone, B. J. Qualheim, R. L. Hoff, and Albert Ghiorso.

Associated research using mendelevium-256 produced in bombarded einsteinium targets, performed by Kenneth Hulet, R. W. Lougheed, J. D. Brady, R. E. Stone, and M. S. Coops at Livermore and J. Maly and B. B. Cunningham at Berkeley, has demonstrated the existence of a relatively stable II oxidation state in mendelevium. This interests me very much because I predicted the existence of such an oxidation state many years ago as a consequence of the actinide concept for the electronic structure of the heaviest chemical elements. The details of both these findings have been submitted for publication in scientific journals. In my opinion, these results lead to the prediction of a quite stable II oxidation state in element 102.

The mention of two of my former students reminds me of how important the Lawrence Radiation Laboratory and the Department of Chemistry, Berkeley, have been in training talented nuclear scientists not only for Livermore but also for laboratories all over the nation and abroad. In my travels for the Atomic Energy Commission, I rarely go into a nuclear installation in this country or abroad without encountering a former student or colleague. I am sure Edwin M. McMillan, director of the laboratory, has had the same experience in the field of physics, as has John H. Lawrence in biomedicine.

I would like to say a word here about a few of my own former

Ernest O. Lawrence, inventor of the cyclotron and founder of the Lawrence Radiation Laboratory, at the control panel of the 60-inch cyclotron shortly after it began operating in 1939. With him is his brother, John H. Lawrence, pioneer in nuclear medicine and former director of the Donner Laboratory, 1948-1969.

Ph.D. students, numbering a dozen, who have played important roles at Livermore. Kenneth Street, Jr., was one of the first to come to Livermore. He rose to the position of deputy director, providing leadership during a critical period. Now at Berkeley as a professor of chemistry, he continues to be a consultant to Livermore. Roger E. Batzel, who finished his training with me in 1951, has long held major responsibilities here; he is now associate director for chemistry and space reactors. Gary H. Higgins is division leader of the Plowshare program. Another former student, Walter E. Nervik, deputy director of the Radiochemistry Division, played a major role in the planning and design of this excellent facility. Still other former students include Dr. Hoff, mentioned earlier in connection with heavy element research, and Harry G. Hicks, who are section leaders, and group leader Dr. Hulet, whose work I have described. Ralph A. James, Richard M. Lessler, David R. Nethaway, and Jose G. Vidal, who are also important contributors at Livermore, received their degrees under my direction. Finally, there is John W. Gofman, associate director for biomedicine, who, after receiving his degree in chemistry under my supervision during World War II, went on to earn an M.D., entered research in medicine, and four years ago organized the Bio-Medical Division here. I should also mention Peter C. Stevenson, director of the Radiochemistry Division, who did postdoctorate work with me.

There are an additional 15 Ph.D. scientists at Livermore who earned their degrees in chemistry at Berkeley with other professors. All are doing important work here. Included among Isadore Perlman's former students, for example, are Robert H. Goeckermann, who, after a distinguished career, including the associate directorship for nuclear testing, has just accepted a position as Atomic Energy Commission scientific representative in Buenos Aires, Argentina; George W. Barton, Jr., a deputy division leader of radiochemistry; and three group leaders, Harris B. Levy, Manfred Lindner, and Floyd F. Momyer, Jr. Also, E. H. Fleming, a student of Burris Cunningham, is assistant head of the chemistry department.

This new facility provides us with a symbolic starting point to recall 15 years of distinguished work by the Livermore Laboratory. The official date of the laboratory's establishment is Sept. 2, 1952. At that time the late Ernest O. Lawrence and Edward Teller assembled a corporal's guard of amazingly young men, most of whom had only recently received their Ph.D.'s. I recall that they could all crowd around a blackboard, and I know that some of this initial group are in the audience today. Their job was to help improve and diversify the nuclear weapons that are so important to American security and to explore some prospects for applying nuclear energy for peaceful purposes. They were infected with the "gung ho" spirit of Ernest Lawrence, a spirit that has prevailed at Livermore ever since. Their performance has been brilliant, imaginative, and invaluable to the nation in both general areas of their responsibility.

In the area of national security, one of the laboratory's most famous feats has been the large role it played in the development of the Polaris missile systems for nuclear submarines and in continuing the advancement of such systems—now the Poseidon. Earlier this year it was my privilege to speak at the commissioning of the new nuclear-powered submarine *Sturgeon* at New London, Conn., and I was aware that Livermore had designed and developed this vessel's Subroc missile nuclear payload. The nation also owes to Livermore the nuclear weapons that arm much of its land-based strategic missile deterrent—including the Atlas missile and some of the Titan and Minuteman missiles. This does not by any means exhaust Livermore's contributions to the diversification and advancement of national security systems, whose existence alone, we all hope, will prevent major wars.

Another striking Livermore contribution to the national security is leadership. Three former directors of Livermore have been chosen for positions of great responsibility in the federal government. The first was Herbert F. York, who was called to become the head of the research and development effort of the Department of Defense. York, now professor of physics on the San Diego campus of the University of California, was succeeded by Harold Brown, who is now Secretary of the Air Force. John S. Foster, Jr.—"Johnny" to everybody—succeeded Brown as Director of Defense Research and Engineering. I can say from personal knowledge that these men have amply justified the confidence placed in them and that their performance has significantly enhanced the national respect for the Livermore Laboratory as an institution. I might add that we in Washington are gratified that the tradition of vigorous and imaginative leadership has been continued in the person of the present director, Michael M. May.

Your achievements in expanding the potential peaceful applications of nuclear energy have been no less striking. Livermore was one of the pioneering laboratories in research on nuclear fusion—the reaction responsible for the energy generated in the sun and which

At the Lawrence Radiation Laboratory at Berkeley, March 1962 (left to right), John Foster, Edwin M. McMillan, Glenn T. Seaborg and President John Kennedy.

Imaginative approaches to fusion. . .

Top: ASTRON. Bottom: 2X II facility.

Plowshare to make economic the recovery of natural resources . . .

---

someday may provide the world with unlimited energy. In the invention of imaginative approaches to fusion and in elucidating the behavior of plasmas, an understanding of which is essential to practical thermonuclear power, Livermore has been in the forefront.

The farsighted Plowshare program was crystallized by Livermore scientists, and this laboratory has had scientific responsibility for this project from the beginning. I am convinced that the world will need Plowshare in the future to make economic the recovery of natural resources such as gas, petroleum, and low-grade ores, as well as for giant earth-moving projects that cannot otherwise be done or whose cost is otherwise prohibitive. Today Livermore scientists, in cooperation with industry and the AEC, are developing the Plowshare technologies to which man will turn in the future. One aspect of present programs in which Livermore is making progress is the development of devices and explosives techniques that reduce the radiation released to the environment. Another major undertaking is the biomedical program, addressed to determining the chain of events from the release of man-made radiation to the environment through to the possible exposure of human beings. This program relates not only to the future safe use of Plowshare, but also to the general problem of human safety as man increasingly uses nuclear energy to serve his needs in the generation of electrical power, in medicine, and in other areas.

Your participation in the Department of Applied Science, Davis–Livermore, is an especially gratifying and timely development. The profusion of problems our society faces includes many that are in the province of very sophisticated applied science. The rapid translation of new fundamental knowledge to practical use can be accelerated by professionals who are masters of the arts of both basic science and engineering. The need for excellent scientists in this field has steadily increased, and today's striking opportunities to create tangible benefits for society can provide unusual personal satisfactions to the practitioners of applied science. We are all indebted to Edward Teller and Albert J. Kirschbaum and to Chancellor Emil M. Mrak and the Davis Campus College of Engineering for recognizing this important need and establishing this promising educational program.

In conclusion, I should like to remark upon the significance of this radiochemistry facility for Livermore's future. You began your work at this former Naval Air Base in old, wooden, barrackslike buildings, some of which you still put to good use. You have gradually acquired suitable facilities. The Theoretical and Computation Building significantly extended your capability in those areas. The seven-story Physics Building and the new Bio-Medical Building, both now under construction, are much-needed additions that will enhance the efficiency of those programs. These capital investments represent a vote of confidence in you by the federal government—and especially by the President, the Atomic Energy Commission, the Department of Defense, and the Joint Congressional Committee on Atomic Energy, which is represented here today by California Congressmen Holifield and Hosmer, Congressmen Anderson and Price of Illinois, and the executive director of the committee, John Conway. This investment, in addition, recognizes that in our scientific–technological age our nation needs the talents and facilities mobilized at Livermore to develop and advance the diverse technologies upon which our future welfare depends. ■

Designation of

EBR-1

as a National Historic Landmark

At a celebration of the event later in the day, George W. Beadle, President of the University of Chicago, and Glenn T. Seaborg.

Designation of the EBR-I as a National Historic Landmark, National Reactor Testing Station, Idaho, Aug. 26, 1966

■ We are greatly honored today by the presence of a man who was closely associated with this nation's atomic energy program in its earliest days and who has followed that program with great interest ever since. Among his many responsibilities, previously as a lawmaker and today as our chief executive, has been that of seeing that the atom serves the nation both in the interest of our national security and world peace. Today he believes strongly that the peaceful atom can be a boon to mankind, and he works for and supports measures to further its progress and fulfill its promises. His being with us in Idaho today, and I am sure much of what he will say to us, attests to that belief.

It is my great privilege and pleasure to present to you at this time the President of the United States, Lyndon Baines Johnson.

PRESIDENT JOHNSON

Thank you. Chairman Seaborg; Governor Smylie; Senators Church and Jordan; Congressmen White and Hansen; former Congressman Ralph Harding; Governor Calvin Rampton of Utah; Mr. Chuck Herndon, candidate for governor; Mr. Bill Brunt, candidate for Congress; my friend Chairman Holifield of the Joint Committee on Atomic Energy; Congressman Wayne Aspinall of Colorado; Undersecretary of Interior Carver; your own citizen, the Chairman of the FCC, Mr. Rosel Hyde; Admiral Raborn, former Director of CIA; all public officials; members of Congress; ladies and gentlemen:

When Hernando Cortez returned to Spain after exploring the New World, he recommended to Charles I that a passage to India be opened by digging a canal across the Isthmus of Panama. Charles consulted his advisers and then rejected the recommendation because, as he later explained, "It would be a violation of the biblical injunction: 'What God hath joined together, let no man put asunder.' "

I have often wondered what King Charles would have said if faced with the decision to split the atom. For in that act was not only the putting asunder of a part of creation it also contained the potential for destroying creation itself.

We have come to a place today where hope was born that man would do more with his discovery than unleash destruction in its wake.

On this very spot the United States produced the world's first electricity from nuclear energy.*

Only three years ago plans were announced for the first private nuclear power plant that would be competitive without any government assistance. Since then more than 20 such installations have been announced by public and private utility companies. Orders have been placed for power reactors with a combined capacity of more than 15 million kilowatts—more than enough electric power for the homes of all the people of Idaho and seven other western states.

By 1980 nuclear power units will have a capacity of more than 100 million kilowatts of electrical power—one-fifth of our national capacity at that time.

This energy is to propel the machines of progress; to light our cities and our towns; to fire our factories; to provide new sources of fresh water; and to help us solve the mysteries of outer space as it brightens our life on this planet.

We have moved far to tame for peaceful uses the mighty forces unloosed when the atom was split. And we have only just begun. What happened here merely raised the curtain on a very promising drama in our long journey for a better life.

But there is another, darker, side of the nuclear age that we should never forget. That is the danger of destruction by nuclear weapons.

It is true that these nuclear weapons have deterred war.

It is true that they have helped to check the spread of Communist expansion in much of the world.

It is true that they have permitted our friends to rebuild their nations in freedom.

But uneasy is the peace that wears a nuclear crown. And we cannot be satisfied with a situation in which the world is capable of extinction in a moment of error, or madness, or anger.

I personally can never escape for very long at a time the certain knowledge that such a moment might occur in a world where reason is often a martyr to pride and to ambition. Nor can I fail to remember that whatever the cause—by design or by chance—almost 300 million people would perish in a full-scale nuclear exchange between the East and the West.

This is why we have always been required to show restraint as well as to demonstrate resolve, to be firm but not to walk heavy footed along the brink of war.

---

*Dec. 20, 1951.

The Experimental Breeder Reactor No. I, a view looking downward on the reactor shield and the top of the reactor tank.

At the heart of our concern in the years ahead must be our relationship with the Soviet Union. Both of us possess unimaginable power; our responsibility to the world is heavier than that ever borne by any two nations at any other time in history. Our common interests demand that both of us exercise that responsibility and that we exercise it wisely in the years ahead.

Since 1945 we have opposed Communist efforts to bring about a Communist-dominated world. We did so because our convictions and our interests demanded it, and we shall continue to do so.

But we have never sought war or the destruction of the Soviet Union; indeed, we have sought instead to increase our knowledge

The first use of electric power from atomic energy. The bulbs are lighted by the generator at right which operates on heat from the Experimental Breeder Reactor No. I.

and our understanding of the Russian people, with whom we share a common feeling for life, a love of song and story, and a sense of the land's vast promises.

Our compelling task is this: to search for every possible area of agreement that might conceivably enlarge, no matter how slightly or how slowly, the prospect for cooperation between the United States and the Soviet Union. In the benefits of such cooperation, the whole world would share and so, I think, would both nations.

Common reasons for agreement have not eluded us in the past, and let no one forget that these agreements—arms control and others—have been essential to the overall peace in the world.

In 1963 we signed the limited test ban treaty that has now been joined by almost 100 other countries.

In 1959 the Antarctic Treaty, which restricted activity in this part of the world to peaceful purposes, was signed by the United States and the Soviet Union. It has now been joined by all countries interested in Antarctica.

In 1963 the United Nations unanimously passed a resolution

prohibiting the placing in orbit of weapons of mass destruction.

When I first became President—as almost my first act—I informed Premier Khrushchev that we in the United States intended to reduce the level of our production of fissionable materials, and we hoped that he and the Soviets would do likewise. Premier Khrushchev agreed.

I believe that the Soviets share a genuine desire to enlarge the area of agreement. This summer we have been negotiating with the Soviet Union and other nations a treaty that would limit future activity on celestial bodies to peaceful purposes. This treaty would, for all time, ban weapons of mass destruction not only on celestial bodies but also in orbit around the earth.

Arthur Goldberg, our ambassador to the United Nations, has just informed me that much of the substance of this treaty has already been resolved. Negotiations were originally recessed on August 4 of this year, but the Soviet Government has now indicated its willingness to pursue them again as soon as possible. The Soviet Union has joined with us in requesting that all the countries participating in the negotiations be prepared to resume discussions on the twelfth day of next month. I am confident that with goodwill the remaining issues could be quickly resolved.

We are also seeking agreement on a treaty to prevent the spread of nuclear weapons.

This treaty would bind those who sign it in a pledge to limit the further spread of nuclear weapons and make it possible for all countries to refrain, without fear, from entering the nuclear arms race. It would not guarantee against a nuclear war; it would help to prevent a chain reaction that could consume the living of the earth. I believe that we can find acceptable compromise language on which reasonable men can agree. We just must move ahead, for we—all of us—have a great stake in building peace in this world in which we live.

Peace does not ever come suddenly or swiftly; only war carries that privilege. Peace will not dramatically appear from a single agreement, a single utterance, or a single meeting.

It will be advanced by one small, perhaps imperceptible, gain after another, in which neither the pride nor the prestige of any large power is deemed more important than the fate of the world.

It will come by the gradual growth of common interests, by the increased awareness of shifting dangers and alignments, and by the development of confidence.

Confidence is not folly when nations are strong. And the United

Argonne National Laboratory made the EBR-I possible . . .

States and the Soviet Union are both very strong indeed.

So what is the practical step forward in this direction? I think it is to recognize that while differing principles and differing values may always divide us, they should not, and they must not, deter us from rational acts of common endeavor. The dogmas and the vocabularies of the cold war were enough for one generation. The world must not now flounder in the backwaters of the old and stagnant passions. For our test really is not to prove which interpretation of man's past is correct. Our test is to secure man's future, and our purpose is no longer only to avoid a nuclear war. Our purpose must be a consuming, determined desire to enlarge the peace for all peoples.

This does not mean that we have to become bedfellows. It does not mean that we have to cease competition. But it does mean that we must both want—and work for and long for—that day when "nation shall not lift up sword against nation, neither shall they learn war anymore."

I think those thousands of you who are here today at this most unusual event, at this most unusual place—the National Reactor Testing Station—know, perhaps more than other Americans, just what a great force nuclear energy can be for peace and just how much the liberty- and freedom-loving Americans have that as their number one objective. If we could have our one wish this morning, it would be that infiltration would cease, that bombs would stop falling, and that all men everywhere could live together without fear in peace under a government of their own choosing.

Thank you for the courtesy that you do Mrs. Johnson and me to come here and meet with us.

CHAIRMAN SEABORG

It is a privilege and honor to accept, on behalf of the Atomic Energy Commission, this plaque designating the Experimental Breeder Reactor No. I as a National Historic Landmark.

I accept it also on behalf of the National Reactor Testing Station and the Argonne National Laboratory, whose outstanding scientists and engineers made the EBR-I possible and thereby created this landmark in our Atomic Age as well as in our national history.

On this occasion I would like to give special recognition to Argonne National Laboratory, whose pioneering work in the nuclear field has been responsible for so many of the advances we enjoy today. It is because of the past efforts of this great national laboratory and the work it is carrying on today that so many of the promises of the atom are rapidly being realized.

In accepting this plaque, Secretary Carver, it is my fervent wish, and I know it is the wish of all the members of the Commission, of our colleagues in all our national laboratories and other facilities, and especially of you, Mr. President, that this landmark will symbolize our country's determination to use the power of the atom for peace and human progress. To this end, we hope that the coming years will bring added significance to this designation of the Experimental Breeder Reactor No. I as a National Historic Landmark. ■

Chemistry at BROOKHAVEN

The new Chemistry Building.

At the dedication: Maurice Goldhaber, (center) BNL Director: Richard Dodson, Chairman of the Chemistry Department: and Glenn T. Seaborg.

Dedication of the new Chemistry Building, Brookhaven National Laboratory, Upton, New York, Oct. 14, 1966

■ It is a great pleasure for me to be here at Brookhaven today to help dedicate this fine new Chemistry Building. Having been trained as a chemist, I take particular pleasure in seeing this area of knowledge grow—from the standpoint of the development of new facilities and equipment and the ever-expanding knowledge of the field. As most of you know, the National Academy of Sciences' report, "Chemistry—Opportunities and Needs," more commonly referred to as the Westheimer Report, points out that basic research in chemistry has not received attention and support commensurate with its importance in this Scientific Age. I concur in this conclusion.

In the light of this, it is encouraging to be here today to dedicate a new building devoted to chemical research and to review with you some of the continuing and new work that will be done here. I like to think that these new laboratories in your building are a kind of symbol of accomplishment, a recognition of all the excellent chemical research that has been done here at Brookhaven since the Chemistry Department first got started in 1947. In the early days, as I know from my own visits here, the Chemistry Department consisted largely of Richard Dodson and a handful of colleagues. They were valiantly trying to convert World War I barracks and World War II buildings into chemical laboratories that would serve as an active and functioning part of a new national laboratory devoted

The early Brookhaven chemical laboratories. Their architectural openness served in a subtle way to promote and encourage discussion between scientists.

to peaceful research on atomic energy. That many of these early pioneers are still here is also a fine indication of the spirit of enthusiasm for research that has been developed here over the years. Many others who went elsewhere have returned today to join in this dedication, again a tribute to the spirit inspired in its members by the Brookhaven Chemistry Department.

Working in these so-called temporary buildings was admittedly inconvenient. At the same time, however, it seems to have made for a certain informality and flexibility of attitude that is always the hallmark of an active and productive research establishment. Because of the shortages of laboratory space, many of the scientists here in the Chemistry Department had to work at their desks in what would usually have been used as corridors anywhere else. Despite the inconvenience, this architectural openness served in its subtle way to promote and encourage discussion between scientists about their work as well as many other things. I am reminded somewhat of the agora, or marketplace, of the cities of ancient Greece which served as a place for both business and philosophy.

I am struck by the fact that this kind of arrangement was felt to have been so successful that it has been carried over into the new building to help stimulate scientific communication and interaction as in years past. There are many other excellent features of this new building which our visitors will have a chance to see later today. One of the most impressive features, perhaps, is its least conspicuous, and this is its very reasonable cost per square foot. Dr. Bigeleisen, the rest of you here, and the architects should be congratulated for their skillful design. This building's high functional efficiency not only led to significant economies of construction but also will help make this a highly effective tool for fruitful research in the years to come.

When the Chemistry Department of Brookhaven was started nearly 20 years ago, it was with a hardy and ambitious handful of scientists in the pine woods and Army barracks of a relatively remote section of Long Island, many hours from the amenities of downtown New York by the uncertainties of either public or private transportation.

In those early days of postwar peaceful atomic research, the doorways to whole new fields of investigation had just been opened. Hundreds of new isotopes had been discovered. Nuclear reactors, particle accelerators, and highly sensitive new scientific tools were only then becoming generally available for scientists who wanted to solve many long-standing problems of organic, inorganic, and physical chemistry or to take steps toward their solution in a newer and more effective way. Carbon-14 and tritium, for instance, promised to give an entirely new insight into many problems in organic chemistry and biochemistry by permitting scientists to keep track of specific atoms as they passed through a complex series of reactions. The nuclear chemists and the nuclear physicists could work together to pry deeper into the nuclear heart of matter. The complex chemical effects of radiation on matter were known to be important but were poorly understood. Many chemical problems related to chemical technology had to be solved before nuclear power could help turn the wheels of industry and light the homes of our nation.

In this framework of challenging and important things to be done, the Chemistry Department got started in 1947. I recall in some detail much of the fine work done here between 1955 and 1958, a period when I served as a member of the Visiting Committee of the Department of Chemistry at Brookhaven. The momentum of those early days has continued, and the results of the investigations and the researches then begun or planned are an active and effective

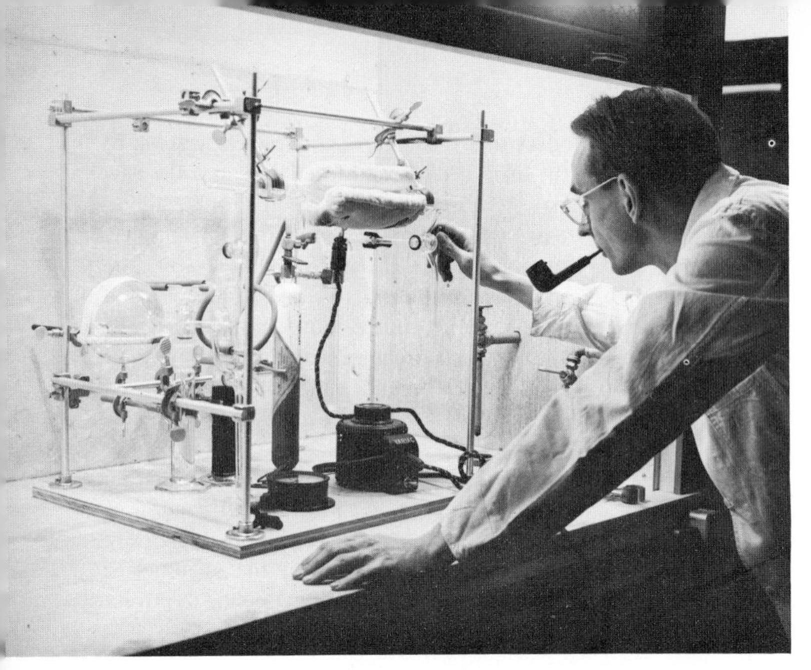

R. C. Anderson investigating a new method for introducing carbon-14 into complex organic molecules, 1948.

part of much of our best contemporary science and technology. Today the laboratories at Brookhaven are well known in all parts of the world. The patterns of interaction between scientists here at Brookhaven and scientists at the universities have continued to be successful. Many graduate and postdoctoral students have worked here, advancing scientific research in areas of advantage to the Atomic Energy Commission and also adding significantly to our country's supply of well-trained and competent scientists. Other scientists and engineers have come here to work for varying lengths of time from industrial and governmental laboratories and from other parts of the world. The work they have done here was useful, and these people have returned to their home organizations with new ideas and capabilities.

If I may, I would like to spend a little time talking about some of the many fine things that have been going on here at Brookhaven in the Chemistry Department. I can touch on only a few of these achievements; after all, just the list of titles of publications from this department since 1947 is some 85 pages long and includes nearly 1100 titles. And this does not include any of the excellent chemical research also done at Brookhaven in other departments, such as basic chemical research done in Nuclear Engineering, in Biology and Medicine, or in Physics.

The nuclear chemists here have effectively exploited the availability of high-energy accelerators—the Cosmotron and the AGS—for detailed studies of the behavior of atomic nuclei under bombardment with fast-moving nuclear particles. Much has been learned about the systematics of these high-energy nuclear reactions, of how the pattern of products formed depends on the choice of target material and the type and energy of projectile. One principal aim of these studies has been to characterize the detailed mechanisms or pathways of the reactions. Another goal is to reach an understanding of the connection between the nuclear reactions observed in complex nuclei and the interactions of elementary particles so intensively studied by the high-energy physicists. Detailed theoretical models and highly advanced computational techniques for the interpretation of high-energy reaction experiments were worked out here in cooperation with scientists at Los Alamos Scientific Laboratory, at the University of Chicago, and at Columbia University. The results of these studies have clarified our knowledge and understanding of the reactions of high-energy particles with complex nuclei. However, they have also shown significant areas in which further improvement and refinement in theory is necessary. This work is currently being pursued. Brookhaven's advanced new computer facilities and staff give strong support to these researches.

This work is closely related to nuclear physics. The skills of the nuclear chemist have been essential for carrying out the chemical separations of the new atoms produced by the bombardment; these separations not only isolate the atoms of interest but also form a vital part of the chain of inference which identifies them. The chemical techniques go hand-in-hand with the study of the characteristic radiations of the radioactive product atoms, which, in turn, characterize the energy levels of the nuclei that emit them. The detailed study of the radiations and energy levels is called nuclear spectroscopy, a field that is effectively pursued at Brookhaven by both chemists and physicists, often working together in the same research effort. The results give a great deal of significant information about the structure and energetics of nuclei.

Nuclear chemistry and spectroscopy can find other unexpected applications, too. The fields of archaeology and the history of art have profited here at Brookhaven because of the imaginative use of nuclear techniques, especially neutron activation, to make highly sensitive but nondestructive chemical analyses of irreplaceably unique works of ancient art. At times this can make for a kind of detective story with clues thousands of years old! This story has been told before but is still worth repeating. Pieces of pottery made

The fields of archaeology and the history of art have profited here at Brookhaven. . .

A technician adjusts a fourth-century glass bottle for X-ray bombardment. The elements in the glass will fluoresce at different wavelengths and reveal its composition. The bottle was discovered in an Italian tomb a century and a half ago. The engraving depicts the luxurious Roman harbor resort of Puteoli, near Naples.

in the ancient city of Arezzo, Italy, were very popular in the Roman Empire because of the high quality of their workmanship. The potters even marked their ware in the same way that manufacturers trademark their goods today. Yet, by means of neutron activation analysis, it has now been possible to show that much of this ancient pottery must have been made elsewhere and given a false trademark, because the chemical composition of the clay is sufficiently distinct to clearly establish that it was not made in Arezzo.

We have mentioned earlier some of the nuclear transformations being studied by scientists here—radioactive decay and induced nuclear reactions. Under ordinary terrestrial circumstances nuclei undergoing such changes are never available in isolated form but are always incorporated in atoms or molecules, which in turn are

Chemist removing a minute sample from an ancient Mayan bowl, in the nondestructive testing of ceramic objects by neutron activation analysis.

surrounded by still other atoms and molecules. It is, therefore, of great intrinsic as well as practical interest to know what effects nuclear processes may have on the chemical systems in which they take place. This field of investigation, sometimes called hot-atom chemistry, has been pursued very actively at Brookhaven in systems ranging from inorganic crystals to gaseous organic compounds. The work on organic substances, principally with highly energetic atoms of radioactive carbon-11 formed in nuclear reactions, has led to new insights into organic reaction mechanisms as well as to potentially useful methods for synthesizing isotopically labelled molecules. Hot-atom studies of inorganic crystals have also proved to be significant in solid-state physics and chemistry.

Related to the chemical effects of nuclear transformations is the broader field of radiation chemistry; that is, the study of chemical changes brought about by energetic radiations, such as X rays, neutrons, gamma rays, electrons, or alpha particles. Through their chemical effects radiations can have manifold practical results, both destructively and beneficially. They can cause the breakdown of an important electronic component in a nuclear control system, or they can be used to sterilize a surgical suture. They can cause a lubricating oil to turn into a hard, brittle solid, or they can help to create a new and more durable plastic. In living matter they can produce a harmful mutation, or they can help to cure a cancer. To understand these complex effects, we must study the basic chemical processes brought about by radiation. In this area of fundamental radiation chemistry, Brookhaven chemists have been among the leaders. In particular, they have long concentrated on the radiation-induced reactions in water and aqueous solvents. Among the important products of water radiolysis is the short-lived and reactive molecular species known as the hydrated electron, which has been intensively studied in the last few years both here and at many other laboratories. Early experimental work at Brookhaven first suggested its existence.

Determinations of kinetic isotope effects have long been used here at Brookhaven to provide insight into the important role of kinetic energy in molecular energy quantization. Isotope effects are an important and established tool in all fields where rates and mechanisms are important, such as in biochemistry and geochemistry. Applications of isotope effects in kinetics and spectroscopy have yielded much interesting and significant information on the mechanisms of a wide variety of chemical reactions and also of the structures of isotopically substituted molecules. In addition, theories of isotope effects find important practical application in the manufacture of enriched isotopic material, such as in the production of deuterium and various kinds of heavy water.

Work on isotope effects here at Brookhaven has always included a strong interaction between theory, experiment and practice, and may confidently be expected to continue on a productive career in the years ahead.

In another important area of modern science, chemists here have been solving difficult problems in the detailed structure of a wide variety of molecules. Modern theories of quantum-mechanical forces and structures are being applied to many complex molecules these days. These highly mathematical theories do much—they both help to build an underlying foundation to all chemical structure and reactivity, and they also help to make life much more difficult these days for graduate students! But these structural theories, like other theories, need to be confronted from time to time with reality in the form of experimental measurements of the exact atomic positions and interatomic forces in the molecules that make up our material world.

Scientists here at Brookhaven are in a highly strategic position; they can use the most powerful modern tools of structure determination, including neutron diffraction, X-ray diffraction, Mössbauer spectroscopy, and advanced computational facilities. The new High Flux Beam Reactor, which has now been in operation for nearly a year, is one of the best tools in the world for neutron diffraction studies, and the use of computer-controlled programs will mean that many new problems can and will be attacked that could not have been tried before. Again, the results of these experiments will be of great interest and value to a large number of scientists. Crystallography remains the best method for the precise and unambiguous determination of complex molecular structures. It is this understanding of the inner architecture of the world of matter that makes the chemist the kind of a scientist he is today and also makes him so valuable a colleague and consultant for the biologist, the medical researcher, the industrialist, and even the geologist and astronomer.

This has been a rather hurried survey of some of the work that has been done here at Brookhaven over the past years and of some of the important things being done today. Perhaps I have emphasized too much the problems and the achievements of the Chemistry Department rather than of the individual scientists, without whom, of course, these accomplishments could not have been possible.

I would like to close with a few words of summary. I think that there have been a great many changes in the science and scope of chemistry in our own lifetime, changes that perhaps have not been fully recognized or understood by the general public, by scientists in other areas, or even by all chemists themselves. Chemists of today are successfully probing the enormous complexities of living matter. They are speculating on the way that life may have first originated here on earth and even on its possible existence elsewhere in the universe. Our world of contemporary abundance is filled with objects newly constructed of the chemist's atoms and molecules—medicines, vitamins, and synthetic substances in great profusion. Chemists are struggling with urgent problems of conservation and purification of water, air, minerals, and our other natural resources. They are helping to solve the problems of food production and preservation that will be ever more urgent in the years to come. In just a few years, chemists will be carrying out detailed analyses of samples from the surface of the moon to get new clues on the origin of the solar system. Chemists have truly moved out of the laboratory into the universe.

But most of all, chemists are extending our basic knowledge and understanding of the physical and chemical world and of its many reactions and changes. If you look in a thesaurus, you will find that one synonym for "changing" is the adjective "protean," derived from the name of the ancient sea god Proteus. In classical mythology Proteus had the power of assuming many different shapes. Menelaus, returning from the Trojan War and searching for necessary informa-

The Electromagnetic Isotope Separator generates beams of ions for research on ion–molecule collision processes.

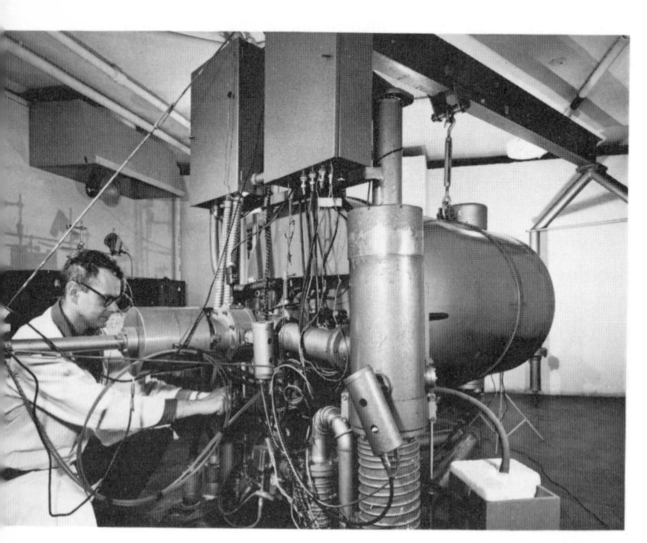

This small Van de Graaff accelerator produces a beam of 2 million-electron-volt electrons for research on radiation effects.

Flow apparatus to study the rates of rapid chemical reactions in solution.

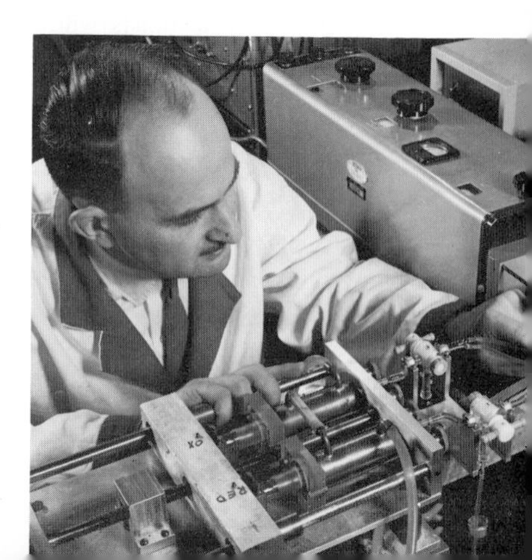

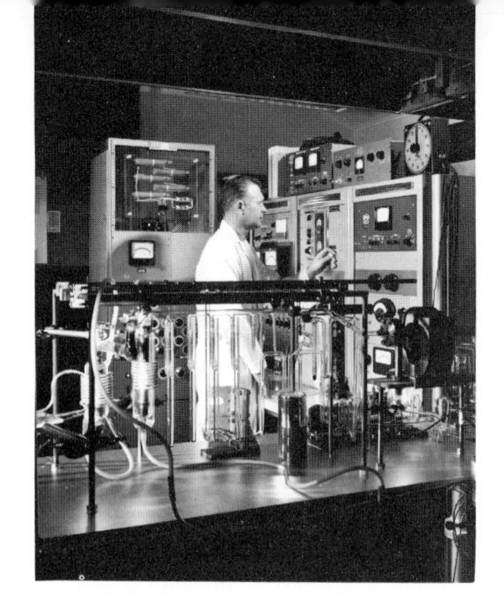

A rare-gas mass spectrometer, which can detect as little as one billionth of a cubic millimeter of gas, used to study meteorite age.

Chemists at Brookhaven are extending our understanding of the physical and chemical world. . .

Jacob Bigeleisen using a cryostat to measure the differences in vapor pressures between isotopic varieties of a substance.

tion, succeeded in grasping the elusive Proteus firmly. Proteus changed first into a lion, then a serpent, then a panther, and even into a stream of running water. Finally, after exhausting his many disguises, Proteus could only return to his original form and then was compelled to give truthful answers to all the questions put to him by Menelaus. We can think of this story as a kind of ancient allegory or fable of the career of the modern chemist, who similarly holds firmly onto matter through all of its manifold protean changes (or energy levels) until at last he gets the answers for which he is searching.

Gentlemen, thank you for inviting me here today. This is an excellent new building in which to do research. I wish you all success in wrestling with Proteus in the years to come. ■

BROOKHAVEN

Brookhaven National Laboratory today.

An Adventure in Scientific Research

Presented at the Twenty-fifth Anniversary of Brookhaven National Laboratory, Oct. 14, 1971

■ The purpose of our gathering today is to commemorate the twenty-fifth anniversary of the founding of Brookhaven National Laboratory and Associated Universities, Inc. Just 25 years ago the governing board of Associated Universities, Inc., launched a new adventure in ways of organizing scientific research. A quarter of a century later there is still that quality of adventure about Brookhaven, that willingness to tackle new problems in new ways, to explore new concepts, and to encourage new talents.

To understand how Brookhaven got started, we must go back to those exciting and rather confusing days in 1945 when World War II suddenly ended with the nuclear attacks on Hiroshima and Nagasaki. Almost overnight those of us who had been working around the clock under unprecedented restrictions of secrecy found our scientific research to be a matter of common concern to people throughout the world. Behind the impenetrable security screen of the Manhattan Project, we had developed a new source of energy that ultimately would touch the lives of all of us.

In August 1945 we had seen how nuclear energy could be used for destructive purposes. The question in the minds of many of us was whether our wartime accomplishments could be directed into channels for human betterment. In the aftermath of Hiroshima, in the chaotic days of readjustment that followed the most devastating war in history, we could not be sure that our hopes could be realized.

It did seem to us who came to be known in the press as "the atomic scientists" that some entirely new approaches to the organization and support of scientific research and development would be necessary. Just as nuclear energy posed some unprecedented and unexpected challenges and opportunities, so we believed old patterns, old ways of doing things would not be sufficient. In fact, one of the problems facing us was whether wartime patterns of research, represented by the laboratories at Argonne, Oak Ridge, and Los Alamos, should continue to dominate the nuclear sciences. Not only was the existing structure inadequate to postwar needs, but also there was some doubt whether the wartime organization could continue to exist even if the government tried to preserve it.

I recall that these fears were very much on our minds in the closing months of 1945. We who were working on transplutonium chemistry at the Metallurgical Laboratory in Chicago faced an uncertain future. Since early 1944 there had been a gradual disintegration of the research staff as many of the scientists who had participated in early studies in physics and chemistry were transferred to other sites. Those of us who remained faced drastic cuts in both personnel and funds as the war drew to a close. Of course many of us were planning to return to our teaching and research in other institutions, but there was still a question of whether we would be able to complete the basic research we had started during the war. When we did return to our prewar positions, we could not be sure that there would be adequate funds and equipment to carry on work in the nuclear sciences. During the war we had come to depend on levels of support which would have been considered fantastic during the lean years of the 1930s.

Finally, there was the spectre of classification. Even if we could find the means for continuing our work, we were not at all certain that our results could be published in scientific journals in the customary way. Obviously, we could not pursue basic research under the kinds of restrictions the Army found necessary to impose upon us during the war.

Some of you probably remember that this concern about the future in the nuclear sciences bubbled over into public protest in the autumn of 1945, especially after the War Department presented its ideas on atomic energy legislation. One center of this protest movement was the Metallurgical Laboratory, where we had been thinking for many months about the postwar implications of atomic energy. I participated in these discussions myself and was one of the signers of the Franck Report. Many of my colleagues in Chicago joined the scientists' lobby in Washington, but some of the most active leaders in the national organization were men who later became prominent scientists at Brookhaven—men like Willie Higginbotham, Clarke Williams, and Lyle Borst.

At the time, I recall, it was rather easy to make the Army, and particularly Gen. Leslie R. Groves, the commanding general of the

Manhattan District, the *bête noir*, the cause of all our problems. As I look back on the situation now, however, it is clear that General Groves and his wartime organization were not antiresearch or antiscientist, but rather they were trying to make the best of a very difficult situation. As a wartime organization under the War Department, the Manhattan District had no clear-cut authority to support activities that would not contribute to the war effort. Although General Groves had to cut back on basic research, particularly after the war ended, he and his assistants were careful not to slam the door completely shut, and they proved receptive to proposals designed to bridge the gap during the transition from wartime to peacetime operations.

When the prolonged Congressional debate in 1946 delayed the passage of the Atomic Energy Act, General Groves and his staff were forced to draw up a budget for the coming fiscal year, including funds for research and development. To assist in this process, Groves appointed an advisory committee consisting of seven scientists who had been prominent in the Manhattan Project. This advisory committee laid down the basic pattern for nuclear research in the postwar period. Foreseeing a substantial expansion in the nuclear sciences, the committee advocated a broad program of basic research on an unclassified basis in universities and private laboratories. The committee also suggested the establishment of "national" laboratories for the primary purpose of pursuing unclassified fundamental research that required equipment too expensive for a university or private laboratory to underwrite. Thus the national laboratories were to be important instruments for implementing government-sponsored nuclear research. The committee suggested that each laboratory have a board of directors chosen from the universities and other participating institutions. The board would submit research proposals and budgets to the government, but some financially responsible and mutually acceptable agency would perform the work of administration.

As a beginning the committee proposed two national laboratories, one at Argonne and the other somewhere in the northeastern states. A planning group at the Metallurgical Laboratory was already at work on the Argonne proposal. The Northeast laboratory was still little more than a hope, but the Manhattan District did set aside $9.4 million for construction of the new laboratory and sufficient funds for operating expenses.

The idea of a Northeast laboratory resulted from some preliminary discussions centered at Columbia University in the autumn of 1945. Among those participating in these early conversations were

A nuclear research laboratory in the Northeast. . .

George B. Pegram, who had been among the first to interest the government in the development of nuclear energy in 1939; Isidor I. Rabi, who headed advanced research for the wartime Radiation Laboratory at the Massachusetts Institute of Technology, and later worked at Los Alamos; and Norman F. Ramsey, the young physicist who had been a key figure at MIT and then at Los Alamos. The discussions had started with the thought of a nuclear research center at Columbia, but the three men soon realized that such a laboratory, including a nuclear reactor and possibly a high-energy accelerator, was beyond the resources of a single institution.

To gather support, Dean Pegram invited twenty other academic and industrial institutions between Philadelphia and New Haven to send representatives to discuss possible arrangements. This meeting, held Jan. 16, 1946, resulted in a letter from Dean Pegram to General Groves proposing the establishment of a nuclear research laboratory in the Northeast. Further meetings of leaders of this group with Manhattan District officials resulted in amalgamation with a similarly interested group in Cambridge, Mass., and in a more detailed proposal in early March sponsored by representatives of six universities and two industrial laboratories. Specifically recommended was the establishment of a laboratory "operated by a single institution as contractor, preferably a single university, but with the scientific direction in the hands of a board representing the sponsoring institutions and appropriate government agencies."

In further discussions it became apparent that sponsorship by several institutions would be more effective, and a decision was made to limit sponsorship to universities. Definitive interest was formally expressed by nine such institutions, Columbia, Cornell, Harvard, Johns Hopkins, MIT, Pennsylvania, Princeton, Rochester, and Yale. At a meeting on March 16, representatives appointed by the university presidents formally joined into an "Initiatory University Group" for the furtherance of interests in establishing and operating the proposed laboratory. This group and its various subcommittees rounded out the plans and brought them to fruition in a remarkably short time.

The sponsors envisioned as the central facility of the new laboratory a large nuclear reactor suitable for physical experiments, the production of isotopes, and the irradiation of reactor components. Such a facility (the only one like it in the world at that time was at Oak Ridge) would require an impressive complex of related chemical, metallurgical, and engineering facilities.

There was also great interest in a high-energy accelerator, but the sponsors decided not to press their luck by asking General Groves for two major research installations.

The third area of special interest to the sponsors was the field of biology and medicine. They recommended that the new laboratory include facilities both for general medical treatment and for research in radiobiology. It is interesting to note that these three disciplines—neutron physics, high-energy physics, and radiobiology—in which Brookhaven has established a worldwide reputation, were clearly evident in the initial planning for the laboratory.

These early discussions also generated some new ideas about the organization of research institutions. These ideas in part explain my earlier statement that Brookhaven from its beginnings has been an adventure in scientific research.

First, there was the proposal to make the laboratory a new type of research institution sponsored by a number of universities. Cooperation between universities was, of course, not a new idea, but the scope and degree of participation proposed for the Northeast laboratory was unprecedented. Certainly it raised difficult problems of organization because it was obvious that representatives of nine universities could not participate in every decision concerning the laboratory. A subcommittee of the Initiatory Group generated the idea of a new autonomous corporation with a board of trustees whose members were drawn from all the participating institutions to serve as the operating contractor with the government. The excellent relations the Commission has had over the years with Associated Universities, Inc. (AUI) have demonstrated the effectiveness of this

---

Neutron physics, high-energy physics, and radiobiology. . .

type of organization. The AUI pattern was adopted in creating the Universities Research Association, Inc., which serves as the operating contractor for the National Accelerator Laboratory, and in a modified form for Argonne Universities Association, which now has a part in the direction of Argonne.

Another novel feature proposed by the founders of Brookhaven was the role of the laboratory as an educational institution. They had no intention of making Brookhaven a kind of super university. The laboratory was not expected to assume from the universities the educational function in the nuclear sciences but rather was to supplement the work of the universities by providing necessary laboratory facilities. For many years this principle has made Brookhaven unique among the Commission's national laboratories. Unlike Argonne or Oak Ridge, Brookhaven over the years has had a relatively small permanent scientific staff. More than any other laboratory Brookhaven has been successful in serving the faculties of the participating universities as a regional research laboratory. In this sense Brookhaven has fulfilled the concept of a national laboratory as a regional center as originally expressed by the Manhattan District's advisory committee on research and development. Indeed, with time and with ease of communication and travel, Brookhaven has become in many areas a national and an international center.

As a part of the cooperative research concept, Brookhaven has maintained close contacts with scientists in all parts of the country. Many have received training at Brookhaven; many others have come here to perform experiments using the laboratory's unique facilities; and others have served as advisors. For a period (between 1955 and 1958) I had the privilege of being a member of the Visiting Committee of the Department of Chemistry, an assignment that gave me a good opportunity to become personally acquainted with the excellent work being done here. Under the able direction of Richard W. Dodson, and more recently under Gerhart Friedlander, the chemistry department at Brookhaven has established an enviable reputation.

The third note of distinction at Brookhaven has been its tradition of independence and of open research. Once the sponsors had submitted their proposal to General Groves in the late winter of 1946, the newly formed association began a series of negotiations with General Groves and his staff. The historical record indicates that these discussions were filled with thorny problems.

General Groves and his assistants saw the new laboratory as a government facility much like Oak Ridge, with all the security restrictions that the scientists at Clinton Laboratories had come to accept. The group of scientists who were establishing Brookhaven had an entirely different concept of the new institution, however; they saw it as being in fact as well as in name a regional research center open to all, with the free inquiry and exchange of ideas which university research teams customarily enjoy. In the opinion of the sponsors, security restrictions on the use of the laboratory equipment and on the publishing of results would defeat the very purposes they were hoping to achieve.

The Manhattan District also assumed that the new laboratory, financed by the government, would be under maximum control and regulation by the government. The sponsors, however, were striving to establish a new type of research institution in which the government would exercise only the most general controls and the actual operation of the laboratory would be in the hands of scientifically oriented management.

The third point of disagreement with the Manhattan District occurred over procedures for constructing accelerators. The Army wanted to put responsibility in the hands of private industry through direct contracts. The sponsoring universities insisted on doing their own accelerator design and development and using industry only as subcontractors. In this disagreement as in the others, the central issue was whether the scientists themselves were to control their own laboratory.

The Brookhaven scientists were not successful in carrying all these points during the early negotiations, but the compromises reached did establish a precedent for an open and independent laboratory operated under policies established largely by the new corporation and financed by the government.

I recall these points of dispute, not for the purpose of digging up old bones of contention, but simply to illustrate the paramount importance that the founders of this laboratory attached to these fundamental principles. They rightly saw these ideas as essential to the operation of a scientific laboratory, and they had the conviction to make these principles part of their new institution.

The Army's Camp Upton was selected. . .

Once these matters had been settled, the founders of the laboratory could turn to more prosaic but equally important questions. Funds, as always, were a problem; in fact, up to this point funds were totally lacking. Administrative personnel were necessary for the laboratory's planning stages, and to pay their salaries a supplement to Columbia University's Manhattan District contract was signed on Apr. 3, 1946. It was soon decided that the new corporation should be established under the laws of the state in which the laboratory was to be located. As a result of this decision, AUI was incorporated in New Jersey on July 8, 1946, since at that time there were more prospective sites in that state than elsewhere. When the present site was selected, incorporation was changed to New York.

One of the original reasons for establishing a national laboratory in the Northeast was that so many of the prospective scientific and technical staff were already located in the region. The sponsors decided that the site for the new laboratory would have to be as near to New York City as possible, be at least ten square miles in area, and have no direct drainage into drinking-water reservoirs (this last requirement was to avoid the possibility of contamination from radioactive waste). After evaluation of several possible locations, the Army's Camp Upton was selected. A training center during World War I and both a reception center and a convalescent hospital in World War II, Camp Upton had attractive features as the site for a future laboratory. Since the entire reservation was scheduled to become surplus property, there would be no restrictions on selection of the area to be developed, and already existing buildings and facilities could be used. The site was near New York City, yet the immediate area was then thinly populated. There was plenty of water, and there was the added advantage that the Long Island sand was an excellent foundation for heavy construction. Use of the existing structures saved at least a year in establishing the laboratory.

By September 1946 a name and a director had been chosen. Philip M. Morse of MIT was appointed director of the Brookhaven National Laboratory, which still existed only on paper and in the

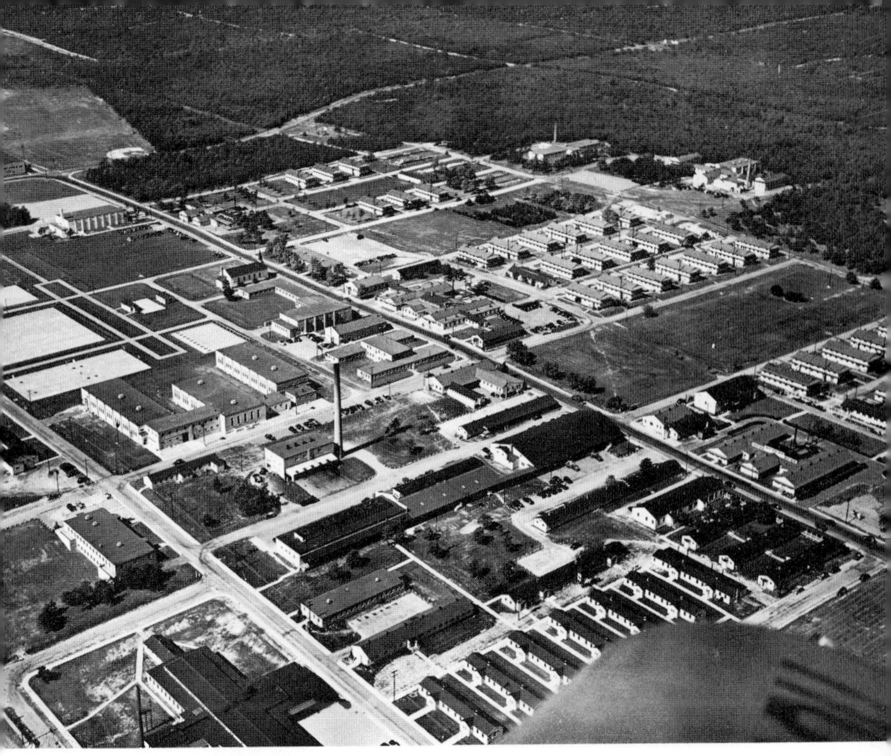

In January 1947, the Army's Camp Upton became Brookhaven National Laboratory as the first scientists began to arrive.

minds of its sponsors. Contract negotiations were in progress with the Manhattan District, but, with the creation of the Atomic Energy Commission, AUI had to deal with the new agency. On Jan. 7, 1947, the Commission issued a letter contract for operation of Brookhaven by AUI, and on January 31 a definitive contract was signed by the two parties.

The first scientists began to move onto the Brookhaven site in January 1947. By the end of the year a staff of about 140 scientists from universities, industrial research laboratories, and other institutions had been appointed, and about 180 scientists had agreed to assist as consultants. Conferences held at Brookhaven in 1947 were attended by nearly a thousand scientists.

The Brookhaven team lost no time in planning a research program and putting it into operation. Perhaps at no point in the brief history of atomic energy had the opportunities for nuclear research appeared brighter than they did in 1947 when work began at Brookhaven. Now, for the first time since 1941, scientists could

A nuclear reactor was a basic requirement. . .

Lyle Borst commences excavation for the Brookhaven Graphite Research Reactor (BGRR), Aug. 11, 1947.

Status of the BGRR in 1948.

The completed BGRR complex in 1949.

A celebration when the BGRR first went critical in 1950. (Leland Haworth in truck).

begin to explore the vast realms of a new physical domain opened up by the discovery of nuclear fission. New ideas and a new laboratory offered the Brookhaven group an extraordinary chance for high adventure, and you made the most of it.

The research program formulated for the laboratory in 1947 was an ambitious one covering a wide range of projects in nuclear physics, including studies of nuclear processes, binding forces, energy levels, and the nature and behavior of fundamental particles. Also to be investigated were the physical, chemical, biological, and medical effects of radiation on all types of matter, including living cells. Another important task was that of seeking a clearer understanding of the physical and chemical properties of artificially produced elements. The application of nuclear materials, such as tracer elements and neutron beams, to nonnuclear areas would also be studied.

A nuclear reactor was a basic requirement for a program of this scope, and, under Lyle Borst and his group, plans quickly developed. Commissioner Sumner Pike broke ground for the reactor in August 1947, and start-up took place in August 1950, a most impressive construction schedule, especially when we realize that one year of rework was required to correct some initial deficiencies. Operation as a research facility began two months later. The Brookhaven Graphite Research Reactor was the world's first reactor designed for peacetime research. During the 1950s scientists from all over the world came to conduct experiments first on the "unclassified face" of the reactor and later with all its facilities. Until its shutdown in 1969, the graphite research reactor was used in a variety of ways and provided much information on atomic nuclei, solid-state structure, and radiation effects. Under the leadership of the late Donald J. Hughes, scientists at Brookhaven produced much of the fundamental data on the nuclear cross sections of the elements, an enduring contribution to nuclear science and technology. The reactor was also an important source of radioisotopes needed for experimental purposes.

Particle accelerators were also vital, and Brookhaven was fortunate to obtain the services of such experienced men as Kenneth Green and John Blewett and, for a time, while they were on leave from their universities, Stanley Livingston and Milton White. Livingston and White were both students of Ernest Lawrence, and I had come to know Green at the University of California in Berkeley where he had been a postdoctoral fellow. Blewett, of course, was an equally distinguished physicist from Princeton. Leland Haworth,

Stanley Livingston in 1933 with the 27-inch cyclotron at the Lawrence Radiation Laboratory, Berkeley.

although heavily involved in management responsibilities, also contributed in detail to the accelerator development program. These men helped to build the high esteem Brookhaven has always enjoyed as a research center for high-energy physics.

The accelerator program concentrated on two fields, relatively low energies under precise control and ultrahigh energies. A 60-inch cyclotron and a Van de Graaff generator would satisfy the lower energy requirements. But for such purposes as multiple meson production and the creation of elementary particles, energies of well over 1 billion electron volts (BeV) would be needed. Careful consideration of various accelerator types indicated that the proton synchrotron would be the most practical device. Although an energy level of 10 BeV was considered feasible, a lower level in the BeV range would mean lower costs, fewer design problems, and earlier completion. Before the end of 1947, the Brookhaven accelerator team settled on a 2.5- to 3-BeV energy level.

By this time I had returned to the Radiation Laboratory at Berkeley, and I watched with great interest the work of Lawrence and his assistants in designing the proton synchrotron there. The Berkeley and the Brookhaven accelerator teams felt a sense of competition since there was the possibility that the Commission would support the construction of only one accelerator. I was also a member of the General Advisory Committee when the Commission

Assembling the Cosmotron, 1950. Workmen are installing a bundle of water-cooled, wound copper bars that form part of the magnet coil. The photograph shows the return winding on the outside of the magnet at the end of a quadrant.

The assembled Cosmotron.

Brookhaven would build what became known as the 3-BeV Cosmotron. . .

asked the committee for a recommendation on accelerator construction. I remember how we decided that two accelerators of different energy levels in the BeV range should be built, one of which should exceed the 5.6-BeV antiproton production threshold, but we could not decide who should build which. We suggested that the two laboratories settle the question between themselves and with the Commission. At a conference at Berkeley in March 1948, the three parties agreed that Brookhaven would build what became known as the 3-BeV "Cosmotron" and Berkeley would build the 6- to 7-BeV "Bevatron." Formal approval for the Cosmotron came in April 1948, and the first construction funds were made available in June.

Celebrating a milestone. Members of the Cosmotron team enjoy a moment of relaxation after a successful test of one quadrant of the magnet, December 1950. G. Kenneth Green stands in the center of the group. From left to right around the circle: Abraham Wise, George G. Collins, Charles H. Keenan, Gerald F. Tape, M. Stanley Livingston, Martin Plotkin, Lyle Smith (mostly hidden), Joseph Logue, and Irving L. Polk.

The magnet foundation was laid in December. Design, construction, and testing of the various components took place concurrently.

March 1952 saw the injection of the first proton beam, and on May 20 the Cosmotron became the world's first accelerator to reach 1 BeV. Commissioner Henry Smyth, one of the "founding fathers" of Brookhaven, dedicated the accelerator on December 15 for the "enlightenment and benefit of mankind." The Cosmotron was retired after it had served high-energy physics for more than 14 years.

Perhaps less dramatic than the construction of a machine capable of accelerating protons to hitherto unattainable energies, but no less important, was the work in radiation biology, a field in which Brookhaven soon took a leading position. Investigations of the response to radiation of a wide range of organisms would be necessary before the implications for human genetics and for other branches of biology could be appreciated.

In 1947 Brookhaven embarked on an ambitious study of plant responses to radiation from both the morphological and the genetic viewpoints and on both gross and microscopic levels. One of the major programs at Brookhaven since 1947 has been the study of radiation effects on plant cells. To my knowledge, Brookhaven was the first to introduce the gamma field, which upon first sight seems a relatively obvious idea. It was, but until 1949 radiation sources were not available for this purpose. This conceptually—and now practically—simple device enabled biologists to observe the effects

Arnold H. Sparrow prepares Trillium *bulbs for irradiation, in 1948.*

of chronic radiation on common food plants, as well as the effects on the cells. Brookhaven's work, led by Arnold Sparrow, has produced a much clearer understanding of the radiosensitivity of various plant species.

Achievement of the BeV energy range was not the only event of significance in the accelerator field at Brookhaven in 1952. The imminent completion of the Cosmotron had attracted an accelerator team from the newly established laboratory of the European Commission for Nuclear Research, called CERN. The CERN group intended to study the Cosmotron as a possible model for its own proposed 10-BeV machine. In the course of reviewing the Cosmotron design to see if it could be extended to 10 BeV, or more, Couránt and Snyder of the Brookhaven staff and Livingston, a summer visitor, discovered the alternating gradient or strong focusing principle. This concept, which was also discovered independently by Nicholas Christofilos in Greece, made possible a reduction in magnet cross section and therefore a reduction in magnet and power-supply costs. It was now economically feasible to design an accelerator with a much larger particle orbit and higher energy.

The CERN delegation was impressed by the concept and, when they returned to Europe, stimulated further study. Both the Brookhaven and the CERN groups began design studies of 25- to 30-BeV alternating-gradient proton synchrotrons in 1953. Collaboration between the two groups was extensive, and the completed accelerators bore a striking similarity. The CERN proton synchrotron was completed in November 1959 and was brought up to 26 BeV. In July 1960 the Brookhaven Alternating Gradient Synchrotron (AGS) reached an energy of somewhat over 30 BeV (33 BeV was eventually achieved).

The alternating-gradient principle made even higher energies possible. In 1967 Russia's 76-BeV Serpukov proton synchrotron replaced the Brookhaven AGS as the highest energy accelerator. The National Accelerator Laboratory's 200-BeV machine utilizes the concept on an even larger scale.

The Brookhaven AGS has been the instrument for discovery of several elementary particles. Among them are the second neutrino (the μ neutrino) discovered in 1962 and the Ω^- hyperon, discovered in February 1964. Another great discovery made at the AGS was the so-called CP noninvariance, which indicates that time-reversal invariance may not hold in some elementary processes. When the present conversion project is completed, the AGS will be an even more versatile tool of high-energy physics.

The Brookhaven Alternating Gradient Synchrotron (AGS). . .

At the dedication: (left to right) Kenneth Green, Emery Van Horn, Leland Haworth, I. I. Rabi, and Maurice Goldhaber.

The AGS and target buildings, a 1971 photograph.

Inside the tunnel of the AGS at the point where protons from the linear accelerator (Linac), entering the tunnel tangentially at lower right, are fed into the orbit of the synchrotron magnet ring.

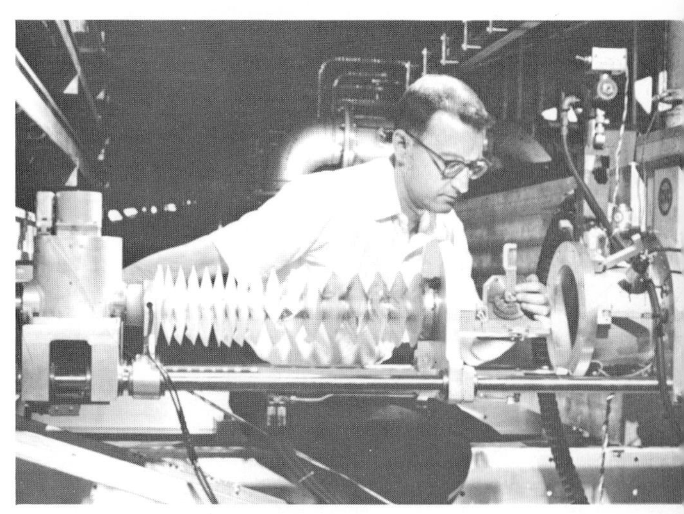

Chemist placing a copper target in the AGS for proton bombardment, in studies on the reactions between high-energy nuclear particles and complex nuclei.

The High Flux Beam Reactor (HFBR). . .

At the dedication in April 1966: (left to right) Maurice Goldhaber, T. Keith Glennan, Gerald Tape, and Emery Van Horn.

The HFBR start-up in November 1965.

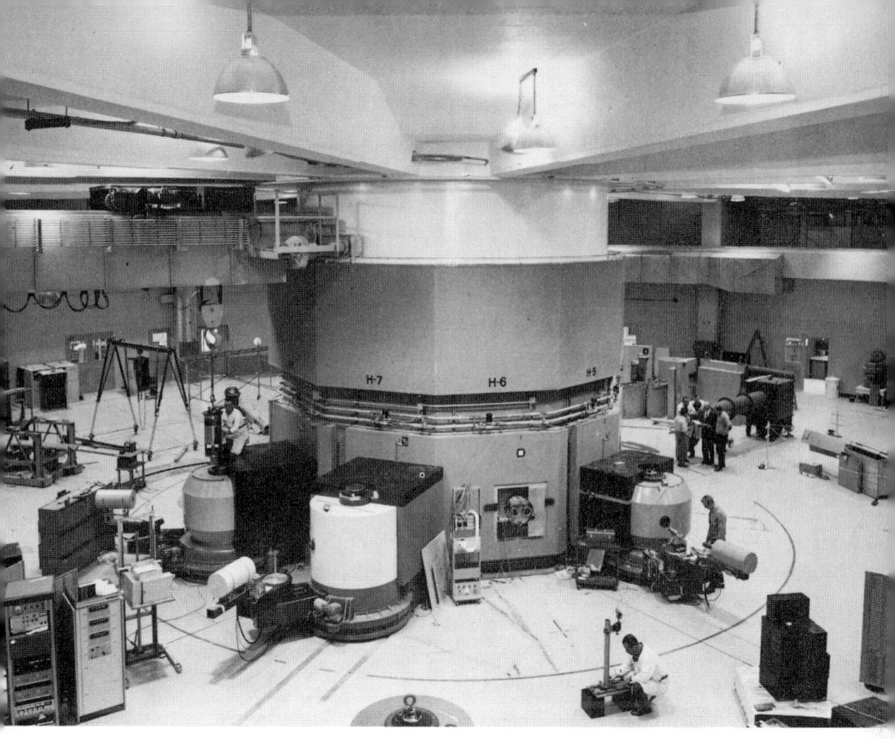

Experimental floor of the HFBR. Three neutron spectrometers, which separate neutrons into various energy ranges, are shown installed at the face of the concrete biological shield to the reactor.

Just as the AGS supplanted the Cosmotron, the High Flux Beam Reactor (HFBR) has replaced the original graphite reactor. This most recent research reactor, which went critical in 1965, is a unique, externally moderated, heavy-water-cooled, -moderated, and -reflected reactor designed primarily for the production of intense neutron beams in the epithermal and thermal energy ranges.

The HFBR has proved remarkably fruitful in its contributions to nuclear physics, solid-state physics, and chemistry. Wide ranging investigations of the structure of solids and molecules, the vibrations of crystal lattices, and the elementary excitations in magnetic substances have been made. The high flux and the specially designed beams and experimental equipment, such as time-of-flight spectrometers and computer-controlled neutron-diffraction equipment, have permitted entry into a new domain of sensitivity and accuracy in this work. Recently, to pick only one example, scientists from Brookhaven, in collaboration with the Bell Telephone Laboratories, demonstrated a new class of magnetic substances in which planes of atoms are independently magnetized. This constitutes a two-dimensional magnetic system, in contrast to previously known ferromagnetic and antiferromagnetic substances whose magnetic structure is three dimensional, and opens the door to a new study of cooperative phenomena in solids.

The team that designed the reactors has made other important contributions to reactor technology and the practical utilization of radiation sources. For example, applied work in reactor physics has led to diverse applications. The importance of the Doppler coefficient as a safety mechanism was first estimated and demonstrated by experiments in the graphite reactor. The theoretical and experimental program on uranium–water lattices, initiated at Brookhaven in 1951, helped pave the way for the design of the pioneering Shippingport and Yankee Atomic Power reactors. The importance of fast fission in ^{238}U and its quantitative measurement was first demonstrated by use of some uranium–water lattices, as was the magnitude of the delayed-neutron yield in the fission of ^{238}U and ^{232}Th.

Closely allied to direct reactor studies are those developments which exploit the by-products of reactors—radioisotopes in sufficient variety to serve the needs of many different sectors of the society. One might also cite the development of a new structural

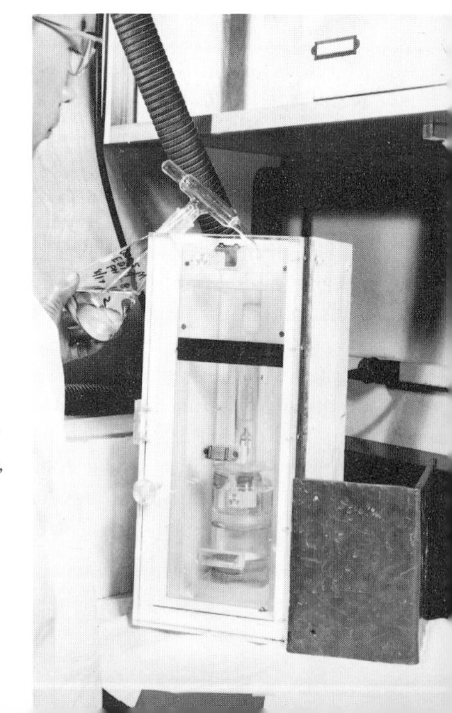

Using a "nuclear cow" to get technetium from its parent isotope. The cow is being fed saltwater through a tube. The saltwater drains through a high-radiation (hot) isotope. The resultant drip-off is a daughter such as technetium-99m. This new, mild isotope can be mixed with other elements, and these become the day's supply of radioisotopes for other scans. Technetium-99m decays with a half-life of 6 hours. Thus greater amounts, with less possibility of injury, can be administered, and a better picture results.

material composed of concrete–polymer produced by the irradiation of concrete saturated with a suitable monomer and the development of those ingenious devices by which short-lived isotopes can be milked from a generator at the point of use in medical diagnostics.

I had the pleasure of reviewing in some detail the outstanding accomplishments of Brookhaven chemists when I spoke at the dedication of your Chemistry Building exactly five years ago today (Oct. 14, 1966).

Reflecting the increased awareness of the impact of technology on the environment, various groups have marshalled ideas and techniques gained over the years as the result of the strong interdisciplinary tradition of Brookhaven. I might mention the clever adaptation of measurements of isotope ratios to the study of the source and atmospheric dispersion of oxides of sulfur from generating-plant stacks.

Brookhaven's research program thus covers a wide range of areas in the physical sciences and technology, but the laboratory is equally well known for its contributions to the biological and medical fields. I have already mentioned the continuing research in plant biology. I would also like to call attention to a few of Brookhaven's more recent achievements in these areas.

The year 1969 was one of special significance both for Brookhaven and for medicine. The Albert Lasker Award for Clinical Medical Research was presented to George Cotzias, head of Brookhaven's physiology division, for his demonstration of the effectiveness of large daily dosages of the amino acid L-dopa in the treatment of Parkinson's disease. Patients treated showed improvement ranging from modest to dramatic. Even modest improvement has meant saving lives in some advanced cases. Treatment with L-dopa reverses chemical abnormalities that are typical of Parkinson's disease. The discovery has been called "the most important contribution to medical therapy of neurological disease in the past 50 years." I am especially proud that a Commission laboratory was the source of this advance.

One more example of Brookhaven's leadership in the application of radioisotopes is the preparation, for the first time in 1956, of tritiated thymidine, a building block of DNA. This labeled substance is of great value in cytological, biochemical, and genetic research. By use of tritiated thymidine, it has been shown that chromosomes labeled during duplication in the cell result in equivalent labels in the chromosomes of daughter cells. A second

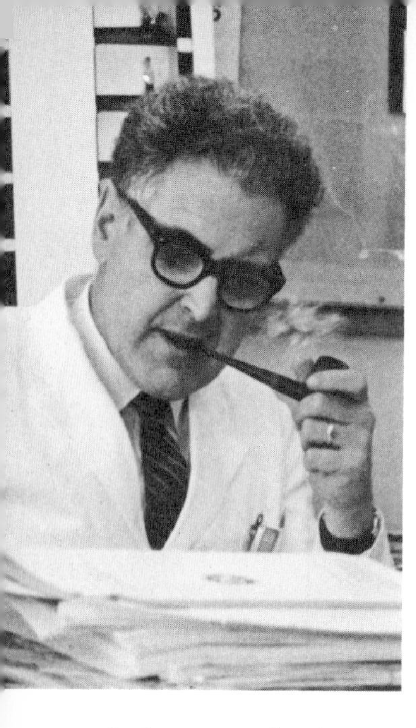

George Cotzias.

duplication in the absence of the labeled material gives a label in only one of the daughter cells. These results confirm the Watson–Crick model of DNA. Using labeled thymidine, investigators in many places have accurately measured the life-span of many different kinds of cells, such as those in the bone marrow, blood, and other tissues, and have systematically examined the proliferative characteristics of cancer and leukemia.

Following another aspect of concern for the environment, ecologists at the laboratory have made part of the natural pine–oak forest on the Brookhaven site the subject of a long-term intensive study on the effects of chronic radiation. Results to date reinforce laboratory studies showing that radiosensitivity is inversely correlated with chromosome volume. Also, in this natural forest radiation appears as a stress much like other stresses, such as that of altitude, in that stress decreases the diversity and therefore the stability of such a complex plant community.

These few examples of the research accomplishments of Brookhaven can in no way do justice to the far-reaching contributions this laboratory has made to nuclear science and technology. I am sure that the achievements of the last quarter century far exceed the most sanguine hopes of those who set out to establish the Northeast laboratory in 1946.

The impact of Brookhaven, however, cannot be measured solely in terms of research reports coming from the laboratory. Perhaps

The irradiation field in 1951, where the effects of chronic radiation on forest plantings are studied.

more important in the long run has been the demonstration that the new type of laboratory proposed in 1946 could succeed, that it could attain the ambitious goals set for it in the days after World War II. The Brookhaven experience has made clear to all that the national laboratory, operating as a truly independent regional research center with government support, has an indispensable place in the structure of modern research institutions. The Brookhaven idea is one of the truly significant inventions to come out of the atomic energy program in the last 25 years.

Its pattern has been used both nationally and internationally. I. I. Rabi, whose wisdom and leadership loomed large in the Brookhaven–AUI endeavor, extended the concept to the Brookhaven of Europe, the present CERN. I would add a personal note of thanks to Rab for his many significant contributions to the Atomic Energy program, as Chairman of the GAC, Scientific Representative to the UN and the IAEA, and in many other roles.

Brookhaven has also had a significant impact upon the Atomic Energy Commission and upon the wider realm of American and world science. I dare not in the short time available today attempt to read the roster of outstanding scientists who have moved on from Brookhaven to other positions of responsibility in the atomic energy

At the banquet celebrating the twenty-fifth anniversary of the founding of Brookhaven National Laboratory, October 13, 1971. From left to right, Glenn T. Seaborg, Isidor I. Rabi, Leland J. Haworth.

program and in the scientific community. I cannot overlook, however, two men who have been especially close to me during my years as chairman of the Commission. Leland Haworth had been director of this laboratory for all but the first year of its existence when I asked him to come to Washington in 1961 as a member of the Commission. Although Lee had become something of an institution himself after 13 years as head of the laboratory, he willingly accepted our call to a higher and in many ways more difficult assignment. After serving two years on the Commission, he accepted President Kennedy's appointment to become Director of the National Science Foundation in July 1963. In six years of

In 1961 at the AGS dedication, (left to right): Gerald Tape, Leland Haworth, and Kenneth Green.

At Brookhaven, October 14, 1971. From left to right, Leland J. Haworth, Glenn T. Seaborg, Gerald F. Tape, Edward J. Brunenkant.

distinguished service in that position, Lee had a profound and lasting effect on the course of science policy in this country.

When Lee went to the National Science Foundation, it was decided that we could do no better than to bring another Brookhaven man to the Commission, in this case Gerald F. Tape, who was then the president of AUI and for many years had been deputy director under Leland Haworth. Jerry became an equally valuable and trusted member of the Commission, and we felt extremely fortunate to have him with us for six years.

I would not wish to suggest by my remarks that all the best people at Brookhaven have left the laboratory for other posts. That is certainly not true. In fact there seems to be a tradition here at Brookhaven for long and distinguished service. Many of the senior staff have spent most of their professional careers in this laboratory. Following Lee Haworth's tenure of 13 years as director, Maurice Goldhaber has just completed his tenth year as director. He, like his predecessors, has had a large part in making Brookhaven the great laboratory it is today.

In conclusion, I think it would be appropriate to say that Brookhaven, perhaps more than any other Commission laboratory, has consistently adhered to the principles upon which it was founded. In an atmosphere free of the restrictions that other laboratories sometimes must live with, Brookhaven has advanced not only nuclear science and technology but also the human condition. Battles in the war against disease have been won, and the structure of the universe has been more deeply penetrated. That quality of adventure which I mentioned earlier will, I am confident, carry Brookhaven to new accomplishments in the years ahead. ■

THE ATOM AT AMES

At the dedication, Frank H. Spedding, director of the Ames Laboratory, and Glenn T. Seaborg in front of the Ames Laboratory Research Reactor.

Dedication of the Ames Laboratory Research Reactor, Iowa State University, Ames, Iowa, May 3, 1963

■ The atom has now been at Ames for a little more than 21 years. I hope that no one regards the new Ames Laboratory Research Reactor as a belated birthday present for your coming of age, for the Ames Laboratory came of age the day it was born and has served our country well ever since.

All of you at Ames are to be commended for your efforts in making the Ames Research Reactor a reality. This fine research tool places the Ames Laboratory on the threshold of a widening spectrum of research opportunities. I have every confidence it will provide the springboard for even greater accomplishments than heretofore attained.

You here at Ames have done much through the years to help the United States maintain its economic and technological superiority. The Ames Project under the Manhattan District played a vital role as a supporting laboratory to the Plutonium Project. Without the basic chemical and metallurgical information developed and supplied by the Ames Laboratory, it is doubtful whether the Plutonium Project would have been completed in time to make its contribution to the revolutionary and awesome nuclear device responsible for terminating World War II.

I know that many of you here today are personally familiar with the truly important contributions of the Ames Laboratory. With your indulgence I would like to recall for the younger people present a few highlights of the early accomplishments of this laboratory and its staff.

The history of the State of Iowa and the present role of the Ames Laboratory in the research and development program of the Atomic Energy Commission represent in important miniature the history and present status of the over-all development of atomic energy for military and peaceful purposes both on this planet and throughout space.

To develop this thesis, I will start by reminding you that the first planet discovered since the passing of the Grecian era was named Uranus by William Herschel in 1781. Soon after its discovery its orbits were computed, and for about forty years the planet could be found at its predicted location in the sky. By 1840, however, the discrepancy between theory and observation had reached distressing proportions. Tackling what was thought by the experts to be an insoluble problem, two capable young men, J. C. Adams, a student at Cambridge, and U. J. Leverrier of France, fired by the cocky optimism of youth, calculated that in order for Uranus to follow its observed path it must be influenced by a new planet in a certain position and orbit. In 1846 (the year that Iowa was admitted to the Union) J. G. Galle, a young German astronomer, found the planet Neptune within one degree of the point indicated by Leverrier.

My reason for mentioning these space discoveries, aside from their great interest, is to remind you that the element uranium was so named by its discoverer, Klaproth, in 1789 as a sort of memorial to the discovery of the planet Uranus by Herschel.

In 1939, exactly 150 years after the discovery of uranium by Klaproth, Otto Hahn and F. Strassmann reported the first definitive experimental evidence for nuclear fission. At about that time Frank Spedding, a young recent graduate of the University of California, was becoming settled in his relatively new position on the faculty here at Ames. Not long after that, in 1940, a new element, named neptunium after the planet Neptune, was discovered by Edwin McMillan and Philip Abelson at the University of California. A few months later the next heavier element, plutonium (also named after a planet—Pluto), was discovered. The study of the nuclear properties of these new elements soon made it evident that plutonium was fissionable, as is uranium-235, and could be produced in large quantities in a chain-reacting unit fueled by the uranium-235 present in natural uranium.

Twenty-one years ago the Nobel laureate Arthur H. Compton assembled leading physicists, chemists, and metallurgists at the University of Chicago to work on the Plutonium Project. The name "Metallurgical Laboratory" was chosen as a cover name to preserve the necessary secrecy. As some of you here today know well, the name may not have been so inappropriate after all. Professor Spedding was asked by Dr. Compton in February 1942 to supervise an important part of the research and planning of that laboratory. He was spending half of each week setting up and directing the chemical research program at the Metallurgical Laboratory and the remainder of the week at Ames directing the correlated research at Iowa State.

At about this time, as some of you will recall, scientists were proposing ways to build nuclear reactors for the production of plutonium, one of the materials required for the atomic bomb development. These reactors were to be fueled by the fission-chain reaction of uranium-235 atoms. The excess neutrons in such reactions were to be used to convert uranium-238 atoms into plutonium-239. This newly discovered element plutonium, which only a few of us had heard of and none of us had seen, had to be generated in kilogram quantities if it was to serve as the basic ingredient in the atomic bomb. It was clear then that tons of highly purified normal uranium metal would be needed for these plutonium producing reactors.

The Metallurgical Laboratory at Chicago did not have the primary responsibility for the procurement of uranium metal. It did seem all important to the success of the Plutonium Program, however, that some action be taken on all phases of the metallurgy of uranium, from obtaining the ore, through purification, production, fabrication, testing, protection, and behavior in use. The responsibility for uranium metal preparation studies, therefore, fell to the group at Ames in February 1942. When small amounts of uranium metal became available from Westinghouse and Metal Hydrides, the Ames research group investigated melting and casting of uranium.

By the fall of 1942, the ingenuity of Frank Spedding and Harley Wilhelm and their colleagues had paid off. Uranium metal produced by this group was eagerly accepted in Chicago as having the necessary quality needed for this important experiment.

The story of the next two months is legendary. As more raw material became available, a production crew was organized, and a metal production plant was authorized to be set up here in the old one-story wooden building (Little Ankeny) left over from World War I. While the conversion of the building and the installation of the equipment were in progress, pilot-plant production of uranium was continued in the chemistry building. More than two tons of uranium ingots were produced in the chemistry building and shipped to Chicago by the end of November 1942.

The world's first chain-reaction pile reached criticality on Dec. 2, 1942. In it were about equal amounts of uranium metal from Westinghouse, Metal Hydrides, and Iowa State College.

When the pilot plant was set up in the old Physical Chemistry Annex, the Ames group continued to produce uranium metal for the Plutonium Project. The production rate by July 1, 1943, had reached several thousand pounds of uranium metal per week. By this time the industrial plants had attained a rate of production that warranted diverting raw material from the Ames Plant.

One of my fond and early recollections of the Ames Laboratory concerns my visit here at about this time. I remember especially the high rate of activity and the dedication and esprit de corps of the scientists and engineers under the leadership of Frank H. Spedding, Harley A. Wilhelm, Adolf F. Voigt, Frederick J. Wolter, William H Sullivan, Amos S. Newton, James C. Warf, and Iral B. Johns, Jr.

The Ames researchers kept the government and the other

Tons of highly purified normal uranium metal would be needed for these plutonium producing reactors. . .

---

laboratories and firms concerned with the development of uranium metal production plants abreast of new developments and assisted them in maintaining the same high quality of metal as that being produced at Ames. Aside from the physical production of badly needed uranium metal at a critical time, the contributions of Ames toward improved quality control and development of increased productive capacity were indeed impressive.

When the war was ended and secrecy was lifted, the Ames Project was awarded the Army–Navy E flag with four stars, signifying two and a half years of excellence in production of metallic uranium as a vital war material under the Manhattan District. The Ames Project

A metal-production plant was set up here in the old one-story wooden building (Little Ankeny) which had been left over from World War I.

(Left to right) Harley A. Wilhelm, A. H. Daane, Amos S. Newton, Adolf F. Voigt, W. H. Keller, C. F. Gray, Frank H. Spedding, R. E. Rundle, and James C. Warf at Ames Laboratory in 1945.

was also mentioned in the original report of the Secretary of War as one of the four outstanding university projects in the atomic bomb program.

Many other important research and development results originated in the Ames Laboratory during the war. As a measure of the activity, more than 100 patent applications were filed on the wartime results.

One of the outstanding characteristics of the Ames program has always been the wholehearted and unselfish cooperation of the entire group. Without this teamwork the success of the project would have been impossible. With this spirit it was clear that the impetus gained during wartime would be carried over with a vigorous postwar program.

Aside from the truly important contributions of the Ames Laboratory to our progress in the World War II atomic energy development, activity here during the war should provide future historians with much raw material for literary development. When Frank Spedding finally sits down to dictate his memoirs, he should be able to write a most interesting document for us. Details of the private life of public men and of the inside and secret history of courts and governments are frequently collected for publication under the title of "Anecdota." Such interesting and amusing

The scientists and engineers under the leadership of Frank H. Spedding. . .

publications, generally of a private nature, are always reported as true. I look forward to receiving my autographed copy of *Spedding's Spiels* in which he will describe in a convincing manner the Hiram Walker whiskey incident. According to my informant, he made out a requisition for 1000 Hiram Walker whiskey barrels but was unable to get the order through the purchasing department. It is said that the order was not cleared until he explained that he actually wanted 1000 empty whiskey barrels to be used as containers for uranium slag, but I have never learned whether he initially hoped to receive the barrels loaded with the product of this well-known company.

One of the most important actions taken by the newly created Atomic Energy Commission in 1947 was its decision to continue the

A meeting in Oak Ridge, Tennessee, Oct. 13, 1947. (Left to right) Stephen Lawroski, John R. Dunning, Frank H. Spedding, Glenn T. Seaborg, and Samuel C. Lind.

The elements rising out of the sixth row are the rare earths.

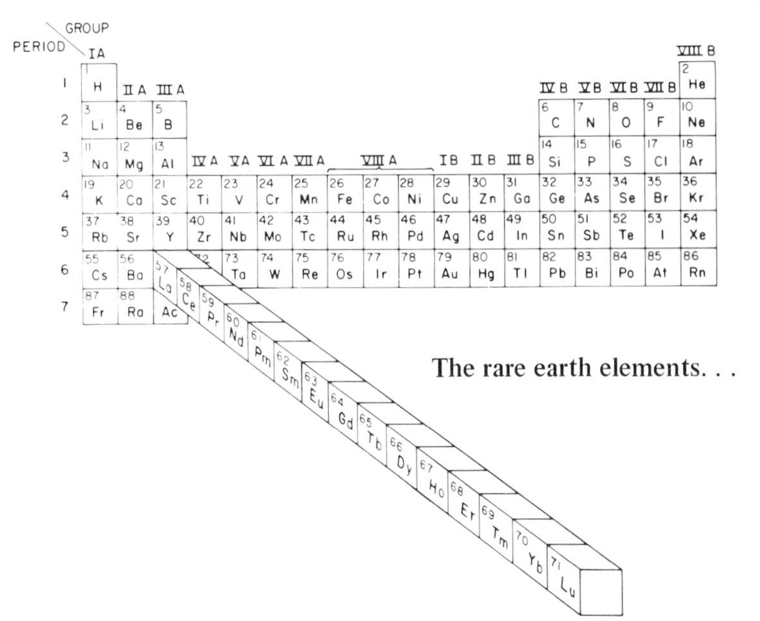

The rare earth elements. . .

Ion-exchange columns used to separate the rare earths from one another. Individual rare earths are collected in the large carboys as they are removed from the columns. The jars in front of the chemist contain the oxides of the individual rare-earth elements.

Ames Laboratory has contributed much to the basic strength of our nation . . .

Ames Laboratory as one of its research installations. Since that time the Ames Laboratory, under the wise and progressive leadership of Frank Spedding, ably assisted by his fine and loyal staff, has contributed much to the basic strength of our nation in the natural sciences.

During his graduate-school days at the University of California, Dr. Spedding became especially interested in the properties of the rare-earth elements and in understanding their baffling behavior. After he transferred his affections to this fine institution at Ames, it was only natural that the intellectual group that grew up around him would decide that the rare-earth elements were deserving of special study. Since these elements constitute a considerable proportion of the fission products, they grow in chemical concentration as the fission chain-reaction process proceeds in a reactor. Some of the rare earths have undesirable nuclear properties and actually decrease the efficiency of operation of reactors as they build up their concentration. Thus an understanding of these materials became highly important.

Recognizing the fundamental importance of knowing exactly what material is under study, the Ames researchers placed emphasis on preparation of the purest possible samples of the rare-earth elements and on knowing, through careful control analyses, precisely what the nature and amount of the remaining impurities are. Today it is fair to say that Ames has, by far, the finest selection of rare-earth elements and compounds of known purity of any place in the world. Distribution of these rare-earth specimens by Ames has facilitated basic research on these elements throughout the world. Scientists and engineers here are also to be congratulated on originating manufacturing processes that are even now being widely used by industry in the manufacture of rare-earth metals.

Here at Ames, too, the physical and chemical properties of the rare-earth elements and other materials are carefully measured. They are measured not only to provide the necessary data for use by engineers in fabricating useful devices but also to understand and explain the basic phenomena of matter, energy, and their interactions.

I have mentioned the work on rare earths because they are the special province of Ames. Splendid research is also under way here in engineering, physics, chemistry, and other fields.

The work of the Ames group in understanding the fundamental behavior of ion-exchange resins has led to many highly useful applications in the separation and purification of many materials, especially the rare-earth elements. The scientific and technical literature abounds with publications attesting to the productivity of the Ames Laboratory in many fields. Some examples that occur to me are methods for separating zirconium and hafnium, methods for separating niobium and tantalum, a technique for separation of uranium-233 from thorium, and liquid-metal extraction methods for removal of fission products and plutonium from power reactor fuels.

The federal and state governments, representative as they are of the citizenry at different levels, must continue to bear increasing responsibility for supporting the educational and research activities of educational institutions. The structures and facilities that comprise this campus are eloquent testimony that the citizens of this state are interested in the pursuit of knowledge. I am hopeful that the addition of this fine new research reactor to the Ames campus will demonstrate to each of you that the federal government has confidence in the staff here at Ames and in the young people who will receive their early scientific training here. ■

MOUND BUILDERS OF THE NUCLEAR AGE

A double anniversary celebration. . .

Mound Laboratory today. The administration building is in the foreground.

At the celebration, Charles A. Thomas, on behalf of the Monsanto Chemical Company, receives a letter of commendation from the Atomic Energy Commission.

Twentieth Anniversary Celebration of the Mound Laboratory, Miamisburg, Ohio, Oct. 18, 1968

■ Being here today in Miamisburg and southern Ohio is especially significant to me personally for two reasons. First, it is not often that I am privileged to take part in celebrating two anniversaries in the same ceremony. The people at Monsanto and the Mound Laboratory who planned this occasion appear to be experts at reducing duplication of effort and consolidating activities. The second and obviously more important reason is that two institutions that have been intimately involved in the growth of nuclear energy are celebrating major anniversaries. Twenty-five years ago the Monsanto Company brought the Nuclear Age to the Dayton area when it accepted the government's invitation to participate in the wartime atomic bomb project. Twenty years ago, back in 1948, when the first building here at this installation was opened for occupancy, the Mound Laboratory became one of the Atomic Energy Commission's first new facilities.

In prehistoric times the Mound Builders of the Ohio Valley, who were highly skilled craftsmen of the Stone Age, fashioned beautiful objects of stone, shell, bone, and beaten metal. In the Nuclear Age highly skilled scientists and technologists began to fashion a new source of energy from matter and knowledge known only to people

Monsanto's work on the chemistry and metallurgy of radioactive polonium-210 became known as the Dayton Project. . .

---

who live in modern times here in the very shadow of a great monument to their Stone Age predecessors.

In the generation that has followed the coming of the Nuclear Age to the Dayton–Miamisburg area, the atom has brought one war to an end and helped to maintain international peace and security. At the same time, the nonmilitary applications of nuclear energy have grown to the point where the atom has become a major force for scientific advancement and technological progress. In this progress in both civilian and defense applications, the Mound Laboratory and Monsanto rank among the leaders in developing the potential of this remarkable and almost unlimited source of energy.

The achievements of companies and laboratories are really the achievements of the people who work in them. So today we are also honoring someone whose contributions to nuclear energy go back to almost the very beginning of the partnership between government and industry in this area of development. Charles Allen Thomas was among the first leaders of industry to influence the course of our nation's nuclear energy policies in both the wartime and peacetime developments. I first became associated with him while I was working on the Manhattan Project some twenty-five years ago. His service to our country is one of Monsanto's major contributions to nuclear energy, and the United States is indeed fortunate that Charles Thomas was one of its early principal statesmen of the atom.

The involvement of Monsanto and Dr. Thomas in the Manhattan Project had an origin that is typical of the way many aspects of the project were organized and launched. During World War II Dr. Thomas was the director of Monsanto's Central Research Department. In 1943 he was called to Washington for a conference with Gen. Leslie R. Groves, the head of the atomic bomb project, and James Conant, at that time a member of the National Defense Research Committee and a leader in this project. After being sworn to secrecy, Dr. Thomas was told about the atomic bomb project. Several days later Monsanto agreed to accept responsibility for an important phase of developing the bomb. Monsanto's task was to work on the chemistry and metallurgy of radioactive polonium-210.

The former Bonebrake Theological Seminary used by Monsanto until Mound Laboratory was completed in 1949.

Another building into which the Dayton Project expanded was the Runnymede Playhouse.

Amidst the urgency of the war effort,
the Dayton Project
had to expand its laboratory facilities. . .

As a source of neutrons, which are produced by (α,n) reactions, for initiating a chain reaction, a purified form of polonium was vital to the construction of the atomic bomb. This work, which was conducted by Monsanto's Central Research Department under Dr. Thomas's leadership, became known as the Dayton Project.

The Dayton Project presented indeed a formidable challenge. At the time no weighable quantities of pure polonium had ever been isolated, and preparing the pure metal required developing some revolutionary scientific techniques. Fortunately the Dayton Project began with some other outstanding scientists who also proved equal to the challenge; among them were James H. Lum, W. C. Fernelius, Malcolm Haring, Joseph Burbage, David Scott, and others.

Not all the obstacles that had to be overcome were scientific. Amidst the urgency of the war effort, the Dayton Project had to expand its laboratory facilities despite some serious shortages of space and equipment. An old building in Dayton which had first been a theological seminary, then a normal school, and then a warehouse was somehow made into a chemical research laboratory. Another building into which Monsanto and the Dayton Project expanded was the Runnymede Playhouse, which had been built on an estate in Oakwood as a recreational facility for the Talbott family. I am sure it was the only scientific laboratory ever to have included a corrugated glass roof, several greenhouses, a stage, lounges, a squash court, an outdoor swimming pool, and an indoor tennis court with a green cork floor.

Despite many problems the Dayton Project made a distinguished record. All commitments for the production of polonium were met, and all shipments were made on schedule. It was not always easy, however, to deliver the purified polonium on time. I understand that sometimes the deadlines were so close that, when a courier arrived to pick up the polonium, some Monsanto employees would have to think of ways to keep him occupied until the final touches were put on the packages being prepared for shipment.

During this time Dr. Thomas was not only directing the polonium project but also was serving the Manhattan Project as a coordinator for closely related work on problems involving chemistry and metallurgy at four different places—Los Alamos, Chicago, Berkeley, and Ames. It was in this role that I first came to know him. I was working to develop processes for separating and purifying plutonium at the University of Chicago's Metallurgical Laboratory. I still recall the meetings he and I attended together and the many conversations we had in which he brought his valuable industrial experience to bear on these chemical and metallurgical problems.

Monsanto's Dayton Project was the immediate predecessor of the Mound Laboratory. Although the war was over, by 1946 it was evident that the nation's need for polonium required a permanent production facility. After being considered along with other sites, the Dayton area, which already had a good supply of scientific talent engaged in polonium work, was chosen. The present location, adjacent to the largest conical Indian mound in Ohio, became the site of the Mound Laboratory.

In the years following World War II, Monsanto continued to make a vital contribution to the nation's nuclear defense effort by operating the Mound Laboratory through a contractual arrangement with the Atomic Energy Commission. Mound rose to prominence first as a manufacturer of nuclear components and also, later, of nonnuclear components related to the development and production of weapons.

In addition to its contributions to military defense, Mound has also shared in the tremendous expansion that has been occurring in the civilian applications of nuclear energy. Today about one-half of the AEC's budget goes for these many and diversified peaceful uses.

One of the most rapidly growing areas in the entire field of nuclear energy is the use of radioactive isotopes. This remarkably versatile form of energy is used in activities related to outer space, medical diagnosis and therapy, industrial quality control, agriculture, food preservation, research, and the development of new products. In the development of isotopic power the Mound Laboratory ranks among the leaders.

Here at Mound, Monsanto has extended the excellence of its wartime work on polonium to the development of this isotope as a peaceful source of nuclear power. During the Manhattan Project, the ingenuity of Monsanto's chemical research led to the development of processes for separating polonium from neutron-irradiated bismuth. Since only a few parts of polonium were mixed with one million parts of bismuth, this was quite a feat. Continuing improvements in producing, separating, and purifying polonium have been made here at Mound. New compounds including a series of rare-earth polonides have been synthesized. The methods developed here have increased processing quantities by a thousandfold and are helping to bring down the cost of polonium to the point where it can be made available as a source of energy where there are special needs, such as space missions that require a minimum of shielding (because it emits very little external radiation) and a high level of thermal power per unit of volume or weight. An example of the usefulness of the alpha-particle radiation of polonium as a source of isotopic power is

The first plutonium-238 source to be orbited in space was the SNAP-3A generator. . .

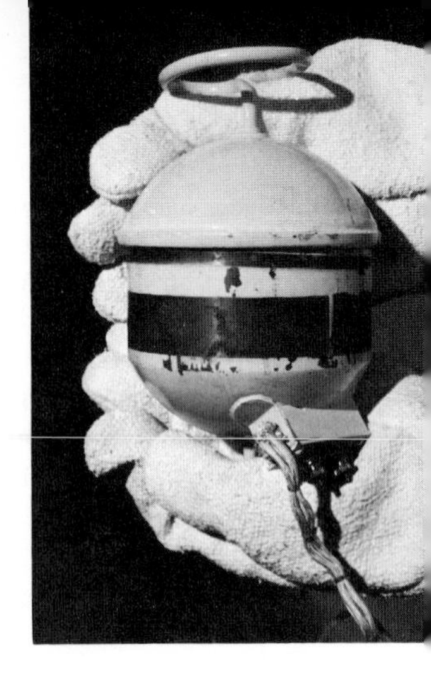

SNAP-3A generator launched in 1961 is still supplying 3 watts of electricity to power the transmitters of an orbiting Navy navigational satellite. SNAP-3A is shown here separately and as attached beneath the satellite. The plutonium-238 heat source was prepared by Mound Laboratory.

the SNAP-29 generator, which the AEC is now developing for Department of Defense applications. The polonium-powered SNAP-29 generator, designed to attain a 400-watt power level, is to be used on space missions of three or four months' duration.

Besides the work on polonium, in some ways the most spectacular developments here at Mound are occurring in connection with another isotope, plutonium-238. The Mound Laboratory prepared plutonium-238 metal and oxide for use as fuel in SNAP generators. People here have developed processing techniques that have greatly increased the usefulness of this man-made material as a radioisotopic fuel. Plutonium-238 is not as well known as plutonium-239, which can be used as the explosive ingredient for nuclear weapons and as a fuel for generating electricity in nuclear reactors. But plutonium-238 may become one of our most valuable isotopes. As a power source in remote regions on land and sea and in space, it has several important advantages. Its half-life is 90 years. The heat from its alpha particle radioactive decay can be used as a compact and lightweight source of electricity. Since very little external radiation is emitted from power units fueled with plutonium-238, it can be handled safely and directly by technicians.

The first plutonium-238 power source to be orbited in space was the SNAP-3A generator. This device weighs only five pounds and is as small as a grapefruit. The Navy put it on board an experimental navigational satellite to provide the power for two radio transmitters. The SNAP-3A is still orbiting around the earth. It is now in its eighth year of operation and is still transmitting signals after having traveled around the world about 40,000 times. The success of the SNAP-3A led to the SNAP-9A, two of which have been launched into orbit on navigational satellites. It is noteworthy that the plutonium-238 heat sources for both of these SNAP generators were developed and fabricated here at the Mound Laboratory.

The Mound Laboratory has also developed and fabricated plutonium-238 fuel elements for two other generators for projects of the National Aeronautics and Space Administration. Two SNAP-19 generators will provide 50 watts of power aboard NASA's Nimbus-B weather satellite. Another plutonium-powered system, SNAP-27, will be left on the moon by our astronauts to serve as a source of power for a package of unmanned scientific equipment. For a year afterward this equipment will send highly significant scientific information back to earth.

Mound is also developing other isotopic-powered equipment for space programs, including an energy source for part of the life-support system for manned space flight. Astronauts on long space missions will need a system to supply potable water. The primary requirements for the system are minimal weight and logistic support. The most significant saving in weight can be accomplished by recovering the crew's waste water products and reusing the water. In one day's time each spacecraft crew member will consume food and drink containing 6 pounds of water. In a year-long mission a six-man crew would use—not counting any water for washing—13,000 pounds. To give you some idea of the effort saved by recycling waste water, I should point out that launching this 13,000 pounds of water on an interplanetary mission would require an additional 20 million pounds of booster thrust.

Radioisotopes will be used as an energy source for the waste-management and recycling system. The Monsanto people here at Mound have fabricated plutonium-238 heat sources for a prototype that has already been tested. Designed to supply two men with water during a 30-day mission, it recovered about three-fourths of a pound

Two other Mound generators for NASA. . .

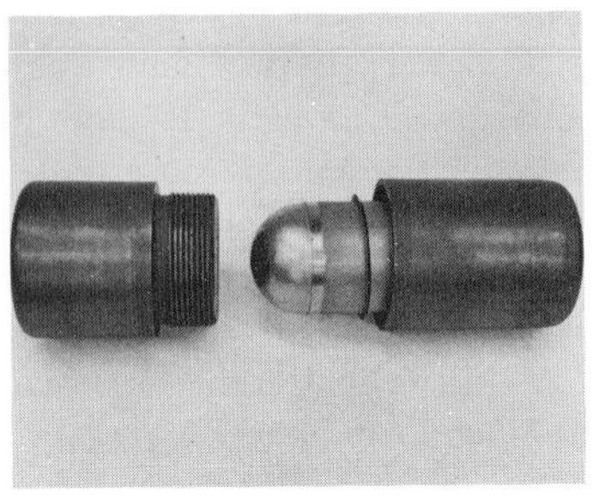

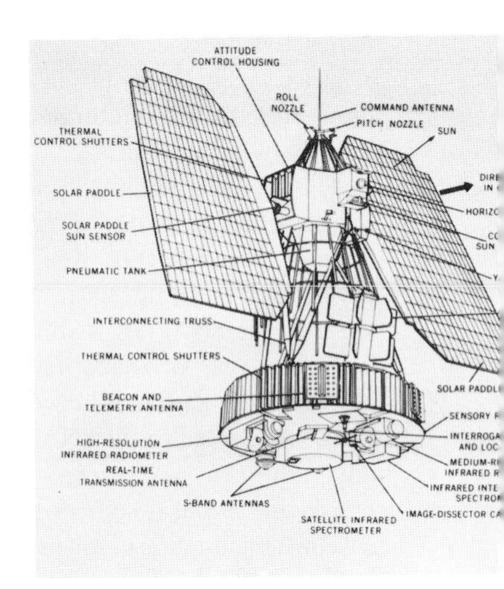

Drawing of Nimbus B weather satellite. SNAP-19 generators are at the center right. Photo at left shows the plutonium-238 dioxide fuel capsule, two of which power the generator. A thermoelectric system converts the heat directly into electricity.

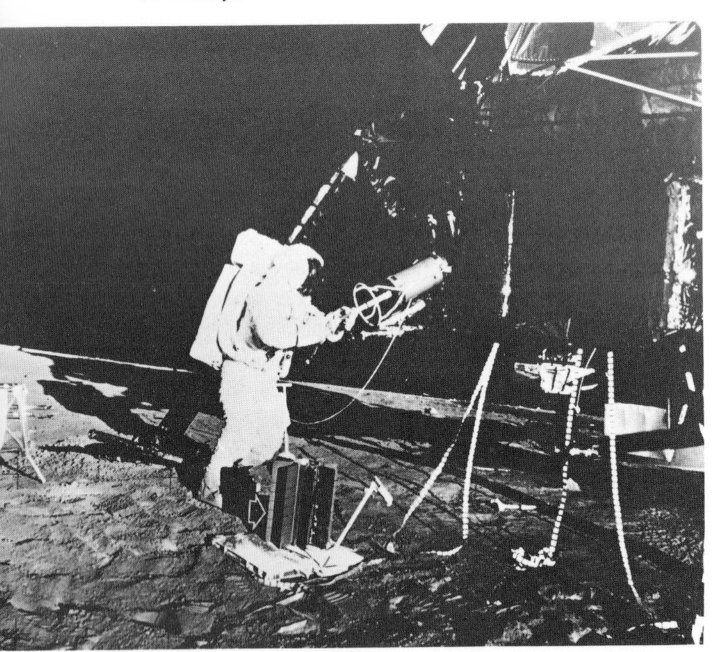

SNAP-27 (arrow), shown on the moon, provides electrical power for the ALSEP instruments left by Apollo 12. Its minimum power output of 63 watts comes from the heat of the radioactive decay of plutonium-238.

of water per hour from waste liquids. Five hundred grams of plutonium-238 were used to supply a total of 280 watts from four isotopic sources for the operation of a rather complicated system consisting of pumps, heaters, evaporators, blowers, valves, condensers, compressors, rotors, and separators.

Besides being involved in some very important and exciting space projects, Mound is developing an isotopic system to be used in exploring another new frontier, the ocean depths. The laboratory is working on a heater for a swimsuit for divers who venture into very cold waters. The need for this kind of equipment has been accelerated by the Navy's underseas activities that require divers to carry out long underwater missions in which these deep-sea pioneers need a high level of mental and motor proficiency. Mound's contribution takes the form of a plutonium-238 source that heats water circulating through plastic tubes interwoven through the diver's swimsuit. The importance of this project, however, goes beyond exploring and developing the seas, for this method of heating could probably be adapted for use by Antarctic researchers, aviators, astronauts, and others who are exposed to extreme cold and need a portable, self-contained heater.

Mound's work in developing the applications of plutonium-238 is also playing a vital role in medical progress. Aspects of two of the most important programs in the medical applications of the atom are being carried out here at Mound. One of these programs is the cardiac pacemaker, and the other is the artificial heart. Mound has worked on developing plutonium-238 power sources for both of these devices. Surgically implanted pacemakers powered by conventional batteries are now used to stimulate malfunctioning hearts in cardiac patients. The power source of a nuclear-powered pacemaker would last many times longer than chemical batteries; so there would be no need for frequent surgical replacement. We hope to develop a nuclear-powered pacemaker that would last ten years or longer.

Many people who succumb to heart disease could be saved if it were somehow possible to replace the diseased organ with an artificial heart. One effort in the field of "spare parts" medicine involves development work on an isotopic-powered artificial heart or blood pump. Such a device might be powered by plutonium-238 or promethium-147. The heat from decaying isotopes would be used by a thermodynamic converter to provide a variable output between 1 and 7 watts to operate the blood pump. The Mound Laboratory has already designed and fabricated two plutonium-238–fueled power sources that are being used in studies to develop the artificial heart.

One of the earliest milliwatt generators developed at Mound Laboratory. The polonium-210 heat source is at the center.

Swimsuit heat exchanger showing two of the four plutonium-238–zirconium alloy sources that each exchanger contains.

Apollo Lunar Radioisotope Heater (about 3 inches high). Two of these, left on the moon by Apollo 11 astronauts, protect seismic instrumentation from the cold of the lunar night. Each heater has enough plutonium-238 dioxide to produce 15 watts of heat.

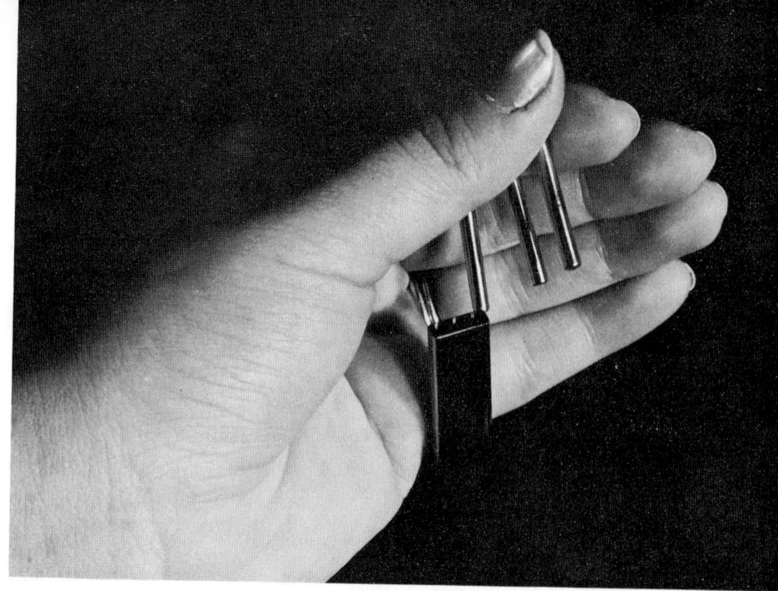

Battery for cardiac pacemaker showing the three plutonium-238 heat sources.

Monsanto's achievements in nuclear energy since World War II, like its vital role in the Manhattan Project, owe a heavy debt to the excellence of its employees. One who exemplifies this excellence and who has carried on in the tradition of Dr. Thomas is the president of the Monsanto Research Corporation, Howard Nason. The AEC is indebted to Mr. Nason not only for his contribution to the program here at Mound but also for his fine work as a member of two AEC advisory committees. He previously served on the Advisory Committee on Isotopes and Radiation Development for four years and is currently on the Atomic Energy Labor–Management Advisory Committee. Today this same tradition of excellence is also being carried on by the director of the Mound Laboratory, Ralph Neubert.

In my remarks today I have tried to highlight some of the accomplishments of the Mound Laboratory and to point out the contribution it continues to make to the diverse and growing field of nuclear energy. Here at Miamisburg, next to one of the earth's undying monuments to the past, the Mound Laboratory stands as a living and working institution for building a better future. Those who have had a part in this—and indeed the whole Miamisburg–Dayton area—can look back proudly on the past and look ahead hopefully to the future. This is both a proud and hopeful anniversary for the Mound–Monsanto–AEC enterprise in nuclear cooperation. I am honored to have had a part in it. ■

THE STANFORD LINEAR ACCELERATOR CENTER

At the dedication of SLAC: (left to right) Congressman Chet Holifield, Glenn T. Seaborg, and Congressman Craig Hosmer.

Dedication of The Stanford Linear Accelerator Center, Stanford, California, Sept. 9, 1967

■ It is a privilege and a pleasure to be here in California with you to dedicate the Stanford Linear Accelerator Center. As many of you know, it is especially gratifying to participate in the dedication of a facility to whose realization one has been able to contribute even in some small measure. When I recall some of my visits here and our many meetings and discussions and when, back in Washington, I look through my enormous file marked "SLAC," I cannot help feeling that I have shared with you some of the birth pains of this important scientific facility. As you may remember, it was a birth with its fair share of travail—some of it having to do with (if I may carry the analogy a bit further) an unusual umbilical cord passing through a nearby community. Even the resolution of that matter carried with it lessons and rewards that I believe will continue to have a beneficial and worthwhile effect.

On this day of dedication, rather than dwelling on our past challenges and accomplishments, I would like to speak to the future—and particularly to the meaning this facility holds for all of us.

The Stanford Linear Accelerator Center is highly significant in many respects and perhaps its primary attraction and significance vary somewhat depending on the interest of each of us. There is no doubt that SLAC is an incredible result of modern science and engineering and a great tribute to its creators. Within this two-mile-long building is housed the world's longest research instrument and certainly one of the most complex and precise machines ever built by man.

In today's science, organized about complex facilities such as we witness here, the men who conceive, design, engineer, construct, and maintain these unique devices contribute to the advancement of science to as great a degree as those who utilize these accelerators to discover new basic phenomena. Great credit is due such men. Foremost among them, of course, is the director of SLAC, Wolfgang K. H. Panofsky. Dr. Panofsky—"Pief" to those privileged to know and work with him closely—has truly been "the father of this facility," having passed through each necessary stage to make this new laboratory a reality. Recognizing the magnitude of SLAC, we can appreciate the burden of his responsibilities. He has been involved in every phase of the development of this mighty research instrument from its very conception about ten years ago to its very

This two-mile long building houses
the world's longest research instrument,
one of the most complex
and precise machines ever built. . .

Subterranean view of Stanford linear accelerator housing. Alignment optics (laser systems) are housed in the large tube, which also acts as support for the smaller accelerator tube above it.

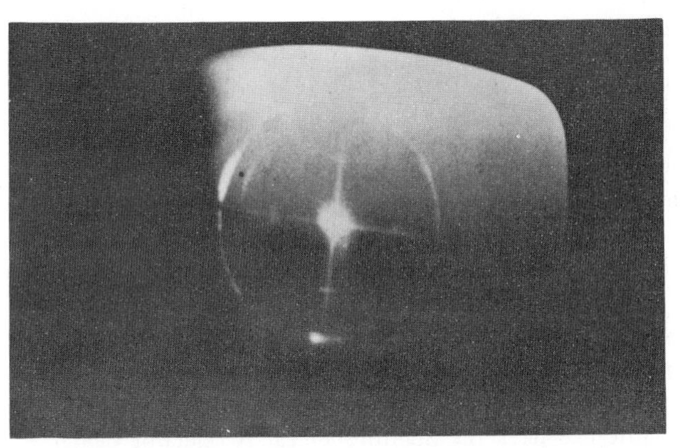

Laser beam spot as observed at the end of the accelerator.

The electron beam originates at the upper end of the 2-mile-long accelerator tube, accelerates continuously as it passes under the 4-lane freeway, and exits at high speed through an underground beam "switchyard" at the end of the tube. Large electromagnets in the switchyard take hold of the beam to deflect it to any of the several well-separated buildings in the foreground.

Wolfgang K. H. Panofsky, professor of physics and director of the Stanford Linear Accelerator Center.

Dr. Panofsky—"Pief" to us—has been involved in every phase of the development of this mighty research instrument. . .

successful initial operations today. He and Deputy Director Matt Sands truly deserve the admiration and gratitude of their colleagues here and of those everywhere engaged in the world of high-energy physics.

Many others have been closely involved in this project from the beginning and deserve our recognition and congratulations. Chief among them is Associate Director Richard Neal. Dr. Neal and his staff are primarily responsible for the working success of the great scientific tool we dedicate today.

What significance does the Stanford Linear Accelerator Center have in the world of science—and in a world where science is playing a central role?

Its significance to the specific discipline of science it was built to serve—high-energy physics—is clear. The energy produced by this accelerator exceeds by a factor of three that of any other electron accelerator in the world today. Its energy is higher by a factor of ten than any other linear accelerator in the world. I might add that

SLAC has already exceeded its design goal by producing more than 20 billion electron volts. Since higher energy implies higher resolution, this means that researchers using this facility will be able to look at the structure of the nucleus of the atom—the protons and neutrons—at a much smaller scale of distance than anybody has been able to do before. Through experiments conducted at this center, physicists will be able to probe deeper into the nature of the nucleus and hence give us greater knowledge of the very basis of matter, energy, and our universe. Experiments in progress here and those yet unformulated will help to expand the frontiers of human knowledge, will contribute to answering some of the great questions that have disturbed the minds of men for centuries, and will lead to the formulation of new and deeper questions. They will intrigue and stimulate men's minds to undertake this newer work that must inevitably follow as we continue to expand the horizons of the human intellect.

Just as we are reaching out into the universe in space for knowledge, so we are delving farther into the universe of the atom. Both quests are essentially for an answer to the same question—one of man's greatest challenges—an understanding of the mysteries of our physical universe. The paths being pursued give all the indications that the laws that govern the large world around us and the inner world of the nucleus are deeply related. Thus the

A linear accelerator. The separation between accelerating gaps, which is the distance traversed by the particles during one half cycle of the applied electric field, becomes greater as the velocity of the particle increases. At any instant adjacent electrodes carry opposite electric potentials. These are reversed each half cycle.

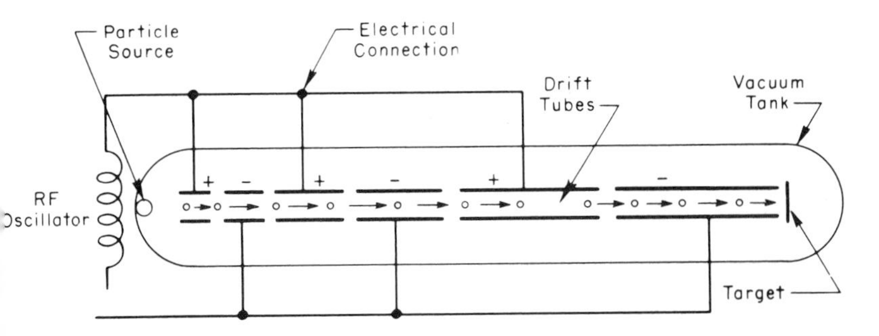

This center will give us greater knowledge of the very basis of matter, energy . . .

---

intellectual significance of SLAC cannot be doubted. It will serve one of man's most fundamental needs; it will further kindle the spark of his curiosity and imagination.

Perhaps I should emphasize at this point that SLAC is open not only to scientists at Stanford University but also to qualified scientists from all parts of the United States and the world. The results of research done here are not classified in any way and will appear in the open scientific literature. Further, both the AEC and SLAC encourage visits to the facility by students and the interested public.

I think it is highly significant that this great research center is open to the international scientific community and that worthy physicists from all over the world may have an opportunity to undertake meaningful scientific research here. The nucleus of the atom is the domain of all men who seek scientific truth. If the combination of this important research tool and the genius of scientific talent—no matter from what nation it may come—results in the revelation of new knowledge, the entire scientific world becomes the benefactor, and ultimately all mankind profits.

What will SLAC mean to Stanford University and to its surrounding community? I believe it will contribute substantially to both. Certainly a university of the caliber of Stanford was duly recognized as one of the nation's great centers of excellence long before ground was broken for this facility. But I think that SLAC and the people who came here to work with it will bring added vigor and stimulation to Stanford. The facility, its personnel, and its many distinguished visitors will enhance the entire community through the center's existence.

I think it is significant that the "cosmic glue" that binds the nucleus of the atom also seems to exert a magneticlike force in attracting men of intellect to a common cause—that of understanding the fundamental forces of the universe and putting these forces to work for man. Perhaps it will be these forces of science, these common quests that men can undertake and work together on and in which they can share the discoveries and benefits, that will in large part be responsible for the fulfillment of one of man's most basic needs—that of being able to live together in peace and understanding.

In conclusion, let me once again offer my congratulations, and those of my fellow commissioners, to all of you who have played a part in creating this unique scientific center. Our best wishes also go to all those who will be at work here, pulsing this center with their own high level of enthusiasm, excitement, and energy.

I understand that during the early days prior to authorization of the project President Sterling told Paul McDaniel, then deputy director of our Research Division, that an underlying condition was "We will not be bound by any precedents or Presidents." I do not believe we have been, and therein may lie the key to the phenomenal success of this project.

The Atomic Energy Commission is proud to be a part of this venture and looks forward to the important accomplishments we know will be forthcoming at the Stanford Linear Accelerator Center. ■

The ARGONNE ACCELERATOR

Instrument and Symbol of Research Progress

Aerial view of the ZGS at time of dedication.

Dedication of the Zero Gradient Synchrotron (ZGS) Facility, Argonne National Laboratory, Argonne, Illinois, Dec. 4, 1963

■ It is indeed a pleasure for me to have this opportunity to participate in these exercises at the Argonne National Laboratory on the occasion of the dedication of the 12.5 billion electron volt Zero Gradient Proton Synchrotron and its associated facilities.

I am especially pleased that all those who were involved with the construction of these high-energy physics facilities have done their work so well. Scientists and engineers from the Argonne National Laboratory, working under the able leadership of Albert Crewe, Roger Hildebrand, and Lee Teng, are to be especially congratulated for the highly professional job they have done to make the ZGS a useful and productive accelerator that scientists from the Midwest can now begin to exploit in their research studies.

The manager of the Chicago Operations Office of the Atomic Energy Commission, Kenneth Dunbar, and his staff can also be proud of their participation in supervising the construction work on the ZGS and in maintaining effective liaison with the many industrial organizations who did so much to make these exercises possible today.

I am also happy to be here where during the past 21 years I have enjoyed close personal and professional relations with many of the scientific, educational, and administrative leaders of this fine institution and of its predecessor, the Metallurgical Laboratory at the University of Chicago. As many of you know, I had the privilege of spending four years in the Metallurgical Laboratory—a period that I recall with nostalgia and much pleasure. I enjoyed my association with many of the people who were members of the laboratory then and who are still making important contributions to the continuing program here. I remember with particular pleasure that many of you working in the divisions headed by Winston Manning and Oliver Simpson came here first at my invitation.

I know that all of you will agree with me when I say that our present harvest of the peaceful benefits of the Atomic Age would not be possible had it not been for the guidance, leadership, and zeal of men like Arthur Compton, Enrico Fermi, Norman Hilberry, and Walter Zinn. We owe a debt of gratitude to these men for their part in laying a firm foundation of scientific excellence for the Argonne National Laboratory. Without such a broad tradition of excellence in chemistry, biology, engineering, reactor development, mathematics, and nuclear physics, I feel certain that the high-energy physics program could not have gotten off to such a fine beginning.

With the successful operation of the Zero Gradient Synchrotron, the Atomic Energy Commission's high-energy physics program has completed an important phase of its work—preparatory, of course, to entering upon a still more important phase, the intensive employment of this facility in tasks of the kind it is uniquely fitted to accomplish. I was in Copenhagen en route to the Seventh General Conference of the International Atomic Energy Agency in Vienna when Dr. Crewe, with emotions resembling those of a new father, wired me the news that the ZGS had been fired up and was in excellent health. Unfortunately the telegram was handed to me as a note that looked like no telegram I had ever seen before. It was only later, in Vienna, that I rescued the crumpled note from my pocket and was immensely pleased by the belated news of AEC's fine new particle accelerator.

There are other interesting aspects of the Zero Gradient Synchrotron's history. You will recall that when the decision was made to authorize the ZGS the 6-BeV Bevatron at the Lawrence Radiation Laboratory at the University of California and the 3-BeV Cosmotron at the Brookhaven National Laboratory had been producing exciting scientific results for several years. Furthermore, at that time the design plans for the 33-BeV Alternating Gradient Synchrotron at Brookhaven National Laboratory were well along, and the Soviets had announced that they would soon be operating a 10-BeV synchrophasotron, as they called it. To complete the picture, the Congress, as of 1956, had authorized the construction of the 3-BeV Pennsylvania–Princeton Proton Accelerator and the 6-BeV Cambridge Electron Accelerator.

All these activities and projected plans were part of a milieu of a decade ago in which it had become clear to many physicists that any real understanding of the nature of the elementary particles and their interactions with each other and with electromagnetic fields would be achieved only by devoting large amounts of human and material resources to the effort. Many curious and baffling events were being observed in high-energy-physics laboratories in this country and elsewhere in the world. To probe into such events with more powerful tools and, if possible, elicit their patterns seemed to both the professional scientist and the layman to be a worthwhile allocation of a significant amount of the resources of this country.

There were those who thought at the time that Argonne National Laboratory should be asked to build a 10-BeV scaled-up version of the Bevatron and have it operating before the Soviet 10-BeV synchrophasotron could be completed. Actually, a project of this kind was authorized for construction in the Midwest in 1956 when it became apparent that the Soviets would be leading the world in maximum proton energy. But the Argonne National Laboratory took a steadier, and what I believe has turned out to be a more productive, view of the situation. After serious study, the laboratory decided that a worthwhile accelerator could not be built within the funds authorized for this project before the Soviets completed their machine. Rather, the laboratory and others recommended that the crash project be cancelled and that the laboratory be authorized to construct a 12.5-BeV Zero Gradient Synchrotron with proton intensities sufficient to carry out the important research experiments envisioned for the accelerator.

That this recommendation, which was approved by the Atomic Energy Commission and by the Congress in 1958, was justified is evidenced by the enthusiasm with which the interested groups of scientists from universities in the Midwest and from the Argonne National Laboratory are approaching the extremely difficult job of assigning priority to the many projects bidding for time in using this facility.

Words of praise for the capabilities of this accelerator have been expressed by many individuals throughout the world. For example, Professor Bernardini, in a speech last spring in Geneva, stated his conviction that the ZGS type of machine had numerous advantages over existing higher energy accelerators in Europe and the United States. As an aside, I might say that representatives of this laboratory, believing as they do in the virtues of rapid communication, lost no time in letting us know in Washington of these laudatory remarks by Professor Bernardini.

(Left to right) Albert V. Crewe, Glenn T. Seaborg, Lee C. Teng, and Roger Hildebrand during a break in the dedication festivities.

There is no good way, of course, to predict the ultimate useful life of a facility such as the ZGS. I am sure all of us would agree that if the ZGS—or any of the other high-energy accelerators now in existence or under construction—should provide us with a clear and definitive understanding of the nature of elementary particles and their role in the structure of matter, it would have earned a rest and could be closed down in good conscience. From present indications, however, there seem to be many years of intensive effort before us until that day is reached. I can only say now, based on the known demands for research time on the ZGS, that this facility—with appropriate improvements—should serve as a very useful instrument for physics in the Midwest for a long, long time.

On an occasion such as this, when we are gathered to share the excitement and satisfaction of realizing this kind of collective scientific achievement, while we are counting our blessings, we should also think seriously of the obligations implicit in the privilege we enjoy of using a major scientific facility of this kind. American science has been fortunate during the last several years in having been given the wholehearted support of the American people. I believe that with wisdom in the goals we select and pursue and with dedicated effort on our part in seeking to achieve these goals, we will continue to merit and receive an even greater degree of intelligent public support. I believe strongly, however, that as responsible scientists we have an obligation to do everything within our power to encourage an intelligent and well-informed climate of public opinion regarding science.

For two reasons this latter comment applies with particular force to the study of high-energy physics. In the first place, it is difficult for the ordinary citizen to see why scientists should make so much fuss about probing into the ultimate secrets of matter. Even though by now the layman may understand that such experiments as these somehow were instrumental in the release of nuclear energy, so much stress has been laid on the short-term benefits to be expected from basic research that the layman may be inclined to expect some beneficial miracle every other week and to show impatience when it is not forthcoming.

Second, his attitude may be even further exaggerated when he learns of the high cost of high-energy physics relative to other fields of basic research. I would emphasize the phrase "relative to other fields of basic research." Even the GAC–PSAC Panel's recommended 10-year program envisaging an annual rate of expenditure rising to 600 million dollars by 1975 for the construction and use of high-energy accelerators is not impressive by comparison with our expenditures on space research and development. It is, nevertheless, a substantial sum.

There is much we can do as individuals, as teachers, and as members of professional groups to help lessen the first difficulty. The citizen in our contemporary world must realize that to participate fully in the life of his age he will have to become much more than the passive recipient of the material dividends of this or that innovation. As one scientist has said, "Education, which transmits from each generation to the next the heritage of the past and the seeds of new powers yet to be, ought . . . to reflect the central reality of modern life." Today's education should, of course, reflect science as being more and more a central reality of our time.

In this context it becomes almost as important for the scientist (including the scientist who deals with the delicate and complex issues of high-energy physics) to provide his fellow citizen with illuminating insights into the broader implications of his discoveries as it is for him to assure that these same discoveries are communicated to fellow scientists with the utmost dispatch and accuracy.

The second difficulty—that concerning the expense of high-energy physics—will also be greatly lessened insofar as the ordinary citizen understands what we are about. He will understand that by their very nature the major investigations of high-energy physics are expensive undertakings that require large and enormously sophisticated machines necessarily manned by large organizations of even more sophisticated scientists, engineers, and technicians. He will understand why, with our continuing demands for ever-higher energies and intensities, the great high-energy-physics research centers are sure to become more, rather than less, expensive.

But when the citizen has reached this desirable level of understanding, he is almost certain to begin asking himself the same questions we are more and more asking each other. I will list a few of these questions—not because any of us presently have the answers, though we may feel we have good first approximations—but because the questions are vital to the sound future development of high-energy physics.

In the first place, our average citizen may raise a question as to the real value of continuing to put ever-increasing emphasis on understanding the role of elementary particles in nature. He may ask if we are choosing an illusory star of hope that will continue to elude our efforts toward some stable definition of its image. Are we making real progress toward our objective? Would we be better advised over the long pull to proceed at a more leisurely pace, or have we already reached a stage in our investigations when it is essential to drive with renewed vigor toward a realizable goal just beyond our grasp? These questions become very important when their answers involve expenditures of the magnitude required to build accelerators, first, in the energy neighborhood of 200 billion electron volts and, later, in the neighborhood of four or five times that figure.

Informally among ourselves and more formally by the members of the GAC–PSAC Panel on High Energy Physics, questions such as these have been discussed and will continue to be discussed. Questions ancillary to these, as well as questions concerning the relation of the programs conducted in high-energy-physics research centers to advanced education and research as undertaken by the universities, must be given our careful consideration.

The conduct of independent research in high-energy-physics facilities as large and as complex as the ZGS raises many issues of fundamental concern to all of us. How can the dedicated scholar continue to participate in the academic and educational life of his university and also, at the same time, make effective use of these large accelerators in his independent research? What are the appropriate ways to make it easier for graduate students from the

Midwest universities and their professors to take advantage of the ZGS and for their counterparts to take advantage of similar accelerators elsewhere?

From an even broader standpoint, what can be done to assure that the ZGS and other high-energy accelerators are made "available to competent scientists and groups of scientists without regard to their current affiliations but, rather, in accordance with the scientific merit of their proposals"? This latter problem assumes crucial importance as we begin to consider retiring older accelerators from service and as more expensive projects push us toward even further emphasis on the research to be conducted at a few very large centers.

Perhaps I have been too serious on such an occasion as this. If so, let me hasten to say that I do not regard any of these problems as insoluble. On the contrary, I believe they are problems that challenge the prospective users of the Zero Gradient Synchrotron to demonstrate as fully as possible its potentialities for very high achievements in the field of high-energy physics. I will be surprised indeed if this new particle accelerator does not become a significant enterprise supported by high-energy physicists and their most promising students throughout the midwestern region of the United States.

With your leave, let me stray from the matter-of-fact for a moment and suggest that even the initials ZGS may, in a fanciful, philosophical sort of way, portend well for this new undertaking. I believe it was the eighteenth century philosopher Johann von Herder who anticipated Spengler and Toynbee in developing the notion that history is a completely determined series of epochs, each fulfilling a specific function and each contributing to the next higher stage. He based this idea on the supposed unique interaction between the spirit or genius of a people and their environment. Von Herder thought this spirit or Zeitgeist of the times determined the moral and intellectual travel of an age or period.

May we let ourselves hope that ZGS can be appropriately transcribed either as Zero Gradient Synchrotron or as *ZeitGeist Scientia.* I believe it is the scientific spirit of our time that will determine both the extent of our travel and the nature of our journey. I believe, also, that the ZGS may prove itself to be a fitting symbol of this spirit and may help lead us to even more rewarding achievements of the intellect during the years ahead.

Let me say once more that I am pleased to see old friends again on this challenging occasion and to participate with all of you in dedicating this new accelerator to the highest service of science and the pursuit of objectives we all cherish. ■

ARGONNE

At the celebration (left to right), Michael V. Nevitt, Congressman Orval Hansen, Robert Duffield, Congressman Melvin Price, Congressman John N. Erlenborn, Philip N. Powers, and Glenn T. Seaborg.

A Tradition of Accomplishment

At the Twenty-fifth Anniversary of Argonne National Laboratory, Argonne, Illinois, June 19, 1971

■ As atomic energy has come of age, there have been several opportunities to celebrate twenty-fifth anniversaries, and I have already taken part in some memorable ones. It is a special pleasure, however, to be with you at Argonne today. I count myself as something of a charter member of the Argonne team. As I shall mention in a moment, I joined the Metallurgical Laboratory long before the idea of the Argonne Laboratory was even thought of. I was still with the Met Lab staff during those months after World War II when the new laboratory was being organized, and I left Chicago only a few days before Argonne came into existence.

The creation of Argonne marked the first attempt in the United States to establish a new type of scientific laboratory, one which would unite in one institution the strong tradition of academic research, which had long been a part of our universities, and the extraordinary advantages of a government-sponsored laboratory, which our experience during World War II had demonstrated. This new institution, called a national laboratory, has emerged in large part from the Argonne experience, and its magnificent accomplishments over the past 25 years have proved the vitality and creativity of this new type of research organization. In this sense the anniversary we are commemorating today has a meaning that goes far beyond the lives of those present and even beyond Argonne itself.

In thinking over the history of Argonne, I recall those exciting days early in World War II when Argonne had its origins in the Metallurgical Laboratory at the University of Chicago. Thanks to the foresight and energy of such men as Vannevar Bush, James Conant, Arthur Compton, and Ernest Lawrence, the United States was ready to launch its effort to build a nuclear weapon when the nation entered the war in December 1941. Within a few days after the attack on Pearl Harbor, Bush and Conant gave Compton responsibility for the research needed to produce a chain reaction and the bomb.

A few weeks later Compton decided he would have to centralize on the Chicago campus much of the research then going on at several universities. Because my group at Berkeley had discovered the element plutonium, which would be the fissionable material produced in the chain reaction, I was invited to Chicago in early February 1942 to discuss our work with Compton, Norman Hilberry, John Wheeler, Enrico Fermi, and others. The Chicago leaders wanted to discuss the production of plutonium and the possibility of devising a chemical method of separating it from uranium and the various fission products of the chain reaction. At this meeting I first fully realized the magnitude of the bomb project and the central importance of our newly discovered element in that enormous effort. I must have appeared confident when I assured Compton that we could develop a separation process for plutonium, but I recall that I had some private misgivings.

Because it would take some time to organize the new laboratory in Chicago and to prepare research facilities, most of the research teams at other universities were scheduled to arrive later in the spring. In the meantime, Fermi and Leo Szilard, with the assistance of Wally Zinn and Herb Anderson, would continue their studies of exponential piles at Columbia. I concluded that my own group would probably stay in Berkeley, where we would be close to the 60-inch cyclotron, which was still our only source of the ultramicroscopic quantities of plutonium we were using in our research. I changed my mind, however, during a luncheon meeting with Norm Hilberry in Berkeley on March 23. I realized that, despite my preference for remaining in Berkeley, I would have to take some of my group to Chicago to develop the separation process.

I will never forget that Sunday afternoon of Apr. 19, 1942, when Isadore Perlman and I stepped off the "City of San Francisco" in Chicago to begin our new adventure. It was my thirtieth birthday, which we celebrated by going to a movie and to dinner in the Loop. Within a few days we were assigned several rooms on the fourth floor of the Jones Chemical Laboratory, which we used as our offices and laboratory. With the arrival of Spofford G. English, one of my graduate students, we had what constituted the entire plutonium chemistry group for more than a month. During these weeks I arrived at the rather novel idea that we might be able to produce enough plutonium-239 by the bombardment of uranium

Spofford G. English, 1945.

Michael Cefola, 1943.

with cyclotron neutrons and the use of ultramicrochemical techniques to study the chemistry of the new element in its pure form. That effort was to demand most of our energies during the spring and summer of 1942.

As the result of two recruiting trips during May and June, I had increased the size of our chemistry group. Michael Cefola from New York University and Louis B. Werner and the late Burris B. Cunningham from Berkeley agreed to join us in Chicago. I also managed to recruit a wife on that Berkeley trip, and Helen returned to Chicago with me to begin married life in a small apartment near the Chicago campus. By that time many other scientists and their families were arriving from universities in all parts of the country. One of the pleasures of being a part of the Met Lab was the opportunity to know and to work with so many people whom we had scarcely seen before. I recall, for example, a picnic Helen and I attended on the Fourth of July weekend in 1942 with the Harrison Browns, the Milton Burtons, and the Perlmans. We went out to the Argonne Forest Preserve to look over the site proposed for the world's first nuclear reactor. Although we had a fine picnic, we never did succeed in finding the reactor site.

During July and the first part of August 1942, the new members of our plutonium chemistry group assembled the specialized equipment for working with extremely small volumes (10^{-5} to 10^{-1} milliliter) and weights (0.1 to 100 micrograms) and developed their techniques with trace quantities of plutonium in microgram amounts of carriers. "Carrier" was the term we used to describe the material which, when precipitated, has the power to sweep out of a solution trace amounts of a desired substance too dilute to be precipitated by itself.

By August 1942 these techniques had been developed to the point where we could attempt to isolate pure compounds of plutonium. After a week of work, Cunningham, Werner, and Cefola finally obtained a solution of pure plutonium compound in a volume of 0.015 milliliters. On August 20 they carefully evaporated this solution until the plutonium concentration became high enough to precipitate as a compound plutonium fluoride. This was man's first sight of plutonium—and in fact of any synthetic element.

As the summer of 1942 waned, the activities of the Met Lab took on a more serious tone. The results of Fermi's research on the critical mass of uranium and our own success in isolating a pure plutonium compound made the idea of developing a nuclear weapon something more than a theoretical possibility. By this time the Army had taken over the project, and we had begun the transition from purely scientific research to engineering development. For our chemistry group that meant planning much larger facilities in the New Chemistry Building on Ingleside Avenue and in a portion of the West Stands. I must admit that for a group of young chemists the idea of the government's spending $200,000 for a building and equipment for our use was an exciting one indeed.

The transition to engineering development caused a similar expansion of thinking in all parts of the laboratory. Some of you may remember that at that time there were tentative plans to build not only the first reactor but also the entire plutonium pilot plant in the Argonne Forest Preserve where we had our July picnic. On Sept. 11, 1942, I again visited this site with Compton, Conant, and other members of the S-1 Executive Committee. I vividly remember Conant's conviction that the site was too close to Chicago for a pilot plant. What we needed, Conant said, was an entirely new perspec-

On Sept. 13, 1942, the famous S-1 Executive Committee which at that time constituted the scientific leadership of the atomic bomb project. (Left to right) Harold C. Urey, Ernest O. Lawrence, James B. Conant, Lyman J. Briggs, Eger V. Murphree, and Arthur H. Compton.

tive; we were, in his opinion, trying to kill elephants with peashooters. As most of you know, the committee then decided that the pilot plant would be built at Oak Ridge.

As it turned out, of course, construction difficulties at the Palos Park site made it impossible to build even the first experimental pile there, and Arthur Compton, with General Groves's support, made the daring decision to initiate the world's first nuclear chain reaction in the heart of Chicago. I well remember the grimy appearance of the workers (some of them are probably here today) who fabricated and assembled the greasy blocks of graphite under the West Stands. On the afternoon of Dec. 2, 1942, that now historic day, I happened to meet Crawford Greenewalt, the young Du Pont executive, in Eckhart Hall just after he had left the West Stands. Greenewalt did not have to say a word to me; I could tell from the glow on his face that Fermi's experiment had succeeded beyond our hopes.

The year 1943 brought a new intensity to our effort to design the plutonium pilot plant to be built at Oak Ridge and the huge production plants ultimately built at Hanford. While Eugene Wigner and others concentrated on the design of the X-10 reactor, we in the plutonium chemistry group were more than preoccupied with the separation process. When we moved into the New Chemistry Building in December 1942, we at last had space to test the various separation processes that had been proposed. Although our knowledge of plutonium chemistry grew at an impressive rate, our research did not indicate that any one process had a clear-cut advantage.

Early in 1943 we decided to use an oxidation–reduction process in aqueous solution, but it was not at all clear whether lanthanum fluoride or bismuth phosphate would be the best carrier of plutonium. Until we made that decision, Du Pont could not fix the design of the Oak Ridge pilot plant. I remember that we discussed the alternatives at a meeting in Chicago on June 1, the deadline Du Pont had established for the decision. Because the engineering data did not indicate a clear choice, Greenewalt turned to me for an opinion. With the fate of the whole wartime project hanging on my judgment, I said I was willing to guarantee at least a 50% recovery of plutonium from the bismuth phosphate process, developed by Stanley G. Thompson of our group. With that assurance, Greenewalt focused most of the engineering talent of his organization on bismuth phosphate. It would be 18 months before I could be certain that my decision had been the right one.

Walter H. Zinn at the CP-2 re[…] controls.

The original West Stands CP-1 reactor, moved from Stagg Field an[…] rebuilt as the CP-2 at the Argonne National Laboratory. Demolished October 1956.

Before the end of 1943, the Oak Ridge pilot plant was in operation, and Du Pont engineers had taken over most of the responsibility for the production plants at Hanford. Supporting work for Hanford and Los Alamos continued, but those of us who remained at the Met Lab also began to turn our attention to the many intriguing possibilities for scientific research which the fission process and the discovery of transuranium elements had opened up. The Palos Park site, which was not used for the first chain reaction, did eventually become the home of the laboratory's experimental reactors—not only of the reconstructed version of the original West Stands CP-1 (then called CP-2), but also of CP-3, the world's first heavy-water-moderated reactor, designed by Wigner and built by Zinn. At this site Zinn also did further studies on fast-neutron reactors and completed the first designs of what was to be the historic Experimental Breeder Reactor No. 1. As the original Met Lab expanded to sites off the Chicago campus, the research facilities at Palos Park took the name of Argonne after the forest preserve. In 1944 Fermi, with Zinn as his assistant, became director of the Argonne Laboratory, which was part of the larger Metallurgical Project under Compton. Thus the now familiar name Argonne Laboratory was born.

Palos Park, May 1947.

The research facilities at Palos Park took the name of Argonne after the forest preserve. . .

In and around the New Chemistry Building. . .

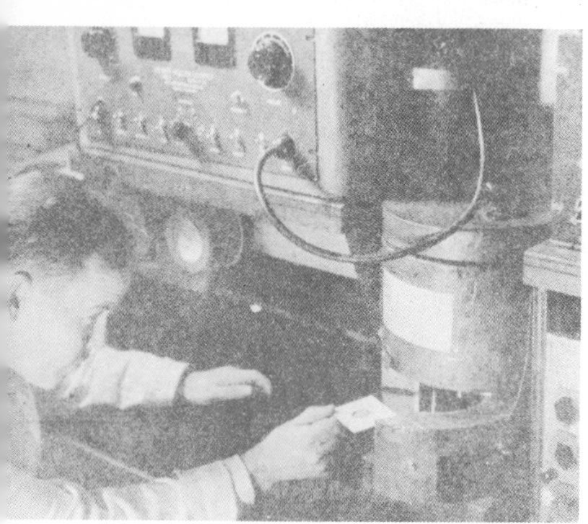

Stanley G. Thompson

Glenn T. Seaborg

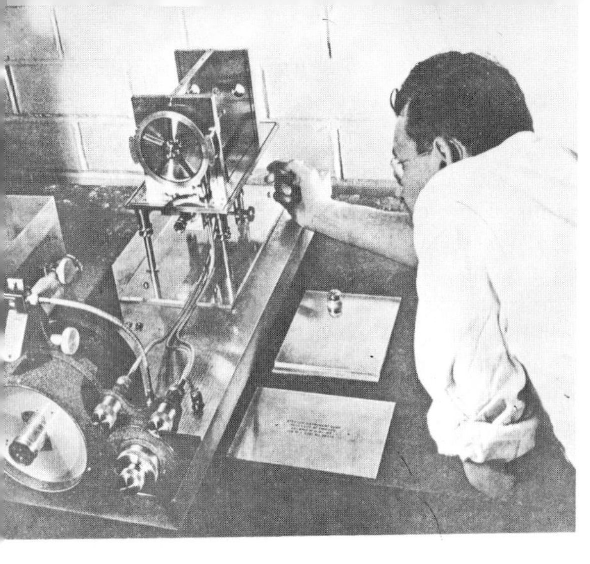

Burris B. Cunningham

(Left to right) Albert Ghiorso, Darrell W. Osborne, Glenn T. Seaborg, James C. Hindman, and Lawrence B. Magnusson.

Thompson and Cunningham.

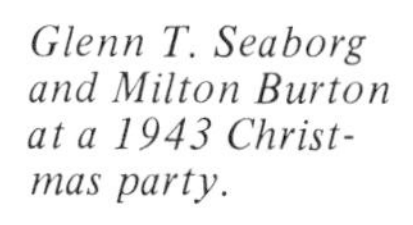

Glenn T. Seaborg and Milton Burton at a 1943 Christmas party.

Those of us still in the chemistry group in 1944 continued our research in "New Chem" with a program that included a search for transplutonium elements. These efforts did not bring any success until we formulated a new theory postulating the existence of a group of "actinide" elements in the heavy-element region with properties similar to the lanthanide rare-earth series in the traditional periodic table. Experiments during the summer and fall of 1944 and extending into the beginning of 1945, using both cyclotron- and reactor-irradiated plutonium, led to the detection of element 96, which we later called curium, and of element 95, which we named americium. During the remainder of the war, in addition to supporting activities at Hanford and Los Alamos, we investigated the processes that made possible the isolation of these new elements in pure form, americium in the fall of 1945 and curium in 1947. As I look back on these events, I realize that some of the most exciting moments of my scientific career occurred in the flimsy laboratories of the Met Lab.

Americium in the fall of 1945
and curium in 1947. . .

Glenn T. Seaborg, lecturing at Northwestern University, Nov. 16, 1945.

The laboratory's rapidly declining responsibilities in 1944 not only made possible some basic research but also forced us to focus some thought on the role we as nuclear scientists might have in the postwar world. At a meeting of the Project Council on Feb. 16, 1944, there was even some discussion of the various types of laboratories that might be engaged in nuclear research after the war. One of these, described as a "cooperative laboratory," should, according to the Council, be established where the scale of research would be "too large to be financed by universities or where secrecy needs could not be met directly by universities." The buildings and equipment would be furnished by the government and research would be administered "by cooperation of educational institutions." This was clearly an early conception of the national laboratory.

These discussions soon led to consideration of the wider social and political implications of nuclear energy. Under the leadership of Zay Jeffries, a laboratory committee set about preparing what Jeffries called a "Prospectus on Nucleonics." Completed in November 1944, the Jeffries report reviewed the possible applications of nuclear science in the near future and the outlook for nuclear power (which seemed good at that time). The committee also recommended that the government support the kind of cooperative laboratories mentioned the previous winter in laboratory meetings. Going beyond the technical aspects of nuclear technology, Jeffries and his committee urged the creation of a world organization to prevent widespread destruction from nuclear war. They also stressed the importance of postwar research in maintaining the U. S. lead in nuclear science and technology.

The Jeffries report had no immediate impact on national policy, but it did help to sensitize many of us at the Met Lab to the difficult policy questions we would be facing as the war ended. This experience made it easier for us to take up the discussion of whether and how to use the first nuclear weapon when that issue came before the Interim Committee in the spring of 1945. Historians may never agree on whether the recommendation of the Franck committee at the Met Lab to provide a demonstration rather than to use the bomb directly ever reached those who made the final decision, but, as a member of that committee, I can assure you that we made a conscientious effort to fulfill our responsibilities as citizens as well as scientists. It was no accident that the Atomic Scientists of Chicago became the leaders in the national debate over postwar atomic energy policy during the summer and fall of 1945.

The "Butterfly Wings" Conference, Chicago, September 1945. At the opening of the University of Chicago's Institute of Nuclear Studies, Samuel K. Allison warned that army security restrictions might force scientists to limit their studies to butterfly wings. (Left to right) seated, S. K. Allison, Enrico Fermi, C. S. Smith, H. C. Urey, and W. H. Zachariasen; standing, C. Eckhart, R. H. Crist, P. W. Schultz, J. E. Mayer, C. Zener, W. H. Zinn, T. R. Hogness, and Edward Teller.

The Met Lab, then, provided a strong and valuable heritage for the new Argonne National Laboratory, which would come into existence in July 1946. First of all, Compton's idea of bringing to Chicago the best available scientists from all parts of the nation created a laboratory on a truly national scale. The Met Lab experience engendered a sense of mission and a standard of excellence every great laboratory must have. Exceptional scientists like Fermi, Wigner, Szilard, and Compton set a pattern of skill, accomplishment, and imagination which we younger scientists tried hard to emulate. That experience trained others like Zinn and Hilberry to carry on the Met Lab tradition and in turn enabled them to impart it to succeeding generations of scientists at Argonne. Furthermore, the concern over postwar policy created a tradition that has inspired Argonne to take a broad perspective in approaching scientific and technical problems. Thus from its very origins Argonne has operated on a principle that others are only now beginning to understand—namely, that the scientist's responsibilities extend far beyond the technical data of the laboratory. These are worthy traditions, and it is to your credit that they are still so much a part of Argonne today.

The idea of a cooperative or national laboratory had taken firm root at the Met Lab since the first months of 1944. The precipitous decline in the laboratory's personnel strength from about 2000 in July 1944 to scarcely more than 1500 in January 1945 caused Compton to recommend that the remnants of the Met Lab be transferred to the University of Chicago, but others, including Zinn, Szilard, Hilberry, and Farrington Daniels, proposed that the laboratory be managed by a board comprised of some 20 universities in the Midwest. The new laboratory would be but one of several "regional cooperative laboratories" that would undertake projects too large for single institutions. They would be financed by the government but would not necessarily be government laboratories.

It is much to the credit of General Groves and his assistant, General Kenneth D. Nichols, that this hope came to fruition in something like its original form. General Nichols sought out representatives of the Midwest universities and asked them to prepare a plan "for continued operation of the Argonne facilities on a cooperative basis between the government and various universities." Nichols then asked the University of Chicago to consider taking over operation of the laboratory on July 1, 1946, "for cooperative research in nucleonics." Argonne National Laboratory came into existence on that date, and Walter Zinn became the first director.

In 1947 the Commission had to rely almost entirely on Zinn and Argonne for its reactor development program. Only one member of the Commission's Washington staff had any reactor experience. The Clinton Laboratories at Oak Ridge had some of the best reactor talent in the nation, but by the spring of 1947 many responsible figures in the atomic energy program doubted that Clinton could survive as a national laboratory. At that time I was a member of the

The first Commission visits Argonne National Laboratory about January 1947, when Walter Zinn (3rd from left) became the first director. Farrington Daniels (5th from left) and members of the first Commission (left to right), Sumner T. Pike, William W. Waymack, Chairman David E. Lilienthal, Robert F. Bacher, and Lewis L. Strauss.

General Advisory Committee (GAC), and I remember that we seriously debated whether, in the face of all the difficulties confronting the Clinton Laboratories, it might not be better to close it down and move the scientific talent elsewhere. In the end, of course, the Oak Ridge Laboratory was saved, but not until the Commission had decided in the closing days of 1947 that it would center all reactor-development work at Argonne.

The enormous responsibility placed upon Zinn and Argonne by this action left little time for the kind of cooperative research in the nuclear sciences which the Board of Governors had contemplated. The Commission had already called on Zinn to draft a reactor development program for the nation, and Argonne was now faced with the task of participating in the design and construction of all but one of the experimental reactors in Zinn's proposal. These included not only the fast-neutron breeder reactor, which Zinn had been developing at the Argonne Forest site, but also two reactors being designed at Oak Ridge. The high-flux testing reactor, the creation of the Clinton Laboratories, would be continued as a joint effort with Argonne. The Clinton scientists and engineers who had been working on a pressurized-water reactor for submarine propulsion moved to Chicago during the summer of 1948, and from that time on Argonne had a major role in developing the propulsion plant for the world's first nuclear-powered submarine.

All these plans for experimental reactors operating at significant power levels raised in a new and serious way the question of finding an adequate site far enough from populated areas to avoid hazards in case of an accident. Zinn and others at Argonne had a key part in discussions that led to the selection of the site for the National Reactor Testing Station (NRTS) in Idaho early in 1949, and the first three reactors built at the Idaho site were in a major sense Argonne products. The Materials Testing Reactor, first operated in 1952, was for more than a decade an indispensable tool for reactor engineers in designing new types of plants and testing components. The Submarine Thermal Reactor, Mark I, which was in operation less than a year later, provided much of the basic technology for pressurized-water reactors.

Leaders in Argonne's submarine reactor program (left to right), Joseph R. Dietrich, Harold Etherington, Wilbert C. Dumprich, Walter H. Zinn, Melvin C. Shaw, and Alfred M. Amorosi.

The Experimental Breeder Reactor No. 1 was uniquely an Argonne creation. . .

The EBR-I.

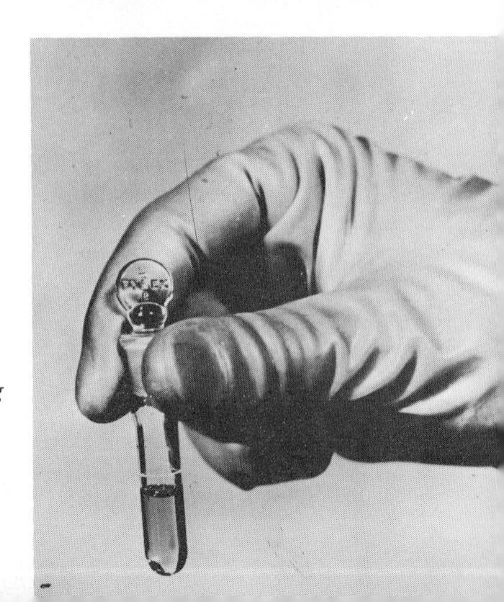

The historic first sample of plutonium produced in EBR-I, which demonstrated for the first time the possibility of breeding fissile material.

The Experimental Breeder Reactor No. 1 was uniquely an Argonne creation and achieved so many firsts in the history of reactor technology, a few of which are listed here. It was the world's first reactor to produce a useful amount of electric power from atomic energy (Dec. 20–21, 1951), the first to demonstrate the possibility of breeding (in 1953), the first to achieve a chain reaction with plutonium instead of uranium as fuel (Nov. 27, 1962), and the first to demonstrate the feasibility of using liquid metals at high temperatures as a reactor coolant.

In addition to this work on experimental power units, Argonne was deeply involved during the early 1950s in developing heavy-water-moderated reactors. Argonne continued to expand its reactor-development activities in the middle and late 1950s. Perhaps of greatest short-run significance was the Experimental Boiling Water Reactor (EBWR), which again was largely a product of Argonne. As the forerunner of numerous full-scale nuclear plants now producing electric power on a commercial basis, the EBWR has a permanent place in the history of reactor development in the United States.

Through most of the 1950s, Argonne, under Zinn's direction, was primarily a center for reactor development, but, by the middle of the decade, new forces were beginning to have an impact on the laboratory. A major force in the changing tides was the growth of Argonne, both in terms of staff and facilities. The scattered buildings of the Met Lab on the Chicago campus and the small warehouselike structures in the Argonne Forest Preserve were now only memories.

The Experimental Boiling Water Reactor, the first nuclear reactor completed in the U. S. Atomic Energy Commission's power reactor program.

Argonne had even moved beyond the temporary Quonset huts the Commission had hastily erected in 1947 to the three separate areas we know today. With an annual operating budget in 1958 of nearly $34 million and a staff of more than 3000, Argonne was attaining physical dimensions and a stature scarcely foreseen a decade earlier. Even more important, the laboratory was no longer heavily concentrated in the reactor sciences, but had grown dramatically in physics, chemistry, and the life sciences. Argonne was now becoming a multidisciplinary laboratory more closely tied to basic research than ever before in its history. Zinn's departure as director in the spring of 1956 was, I think, more a symptom than a cause of the profound changes that were occurring in Argonne. In 1958 the laboratory, under the direction of Zinn's successor, Norman Hilberry, was far more than the Commission's reactor development center, which it had been a decade earlier.

With Hilberry at the helm, this new image of Argonne stimulated within the laboratory long-cherished hopes for new facilities and among the participating universities new demands for a more effective relation. These two interests merged in the long and complicated efforts between 1952 and 1958 to build a new high-energy accelerator, either as a part of Argonne or as the central facility of a new regional laboratory in the Midwest. By the end of that period, the new accelerator was still a dream, but the formation of the Associated Midwest Universities, Inc., made possible closer ties between the laboratory and the neighboring universities.

The decade of the sixties saw a gradual, but major, reorientation of Argonne's reactor program from water reactors to liquid-metal-cooled fast breeder reactors. Shortly after the successful development and operation of EBR-I, as noted earlier, design was begun on EBR-II, an experimental fast breeder reactor power station of 20-Mw(e) capacity whose purpose was to demonstrate the potential technical and economic feasibility of using fast reactors for central-station power plants. This was to be done by both producing electricity and demonstrating the feasibility of the closed fuel cycle.

The closed fuel cycle was a unique feature of the EBR-II. Basically, this amounted to a system whereby fuel was removed from the sodium-cooled reactor and taken apart into its component parts; the fuel sections were treated metallurgically to separate out the plutonium and most of the fission products from the molten uranium; and new fuel, fabricated from the recovered uranium, was reassembled into fuel elements, which were reinserted into the reactor—all this was done by remote control, mostly behind

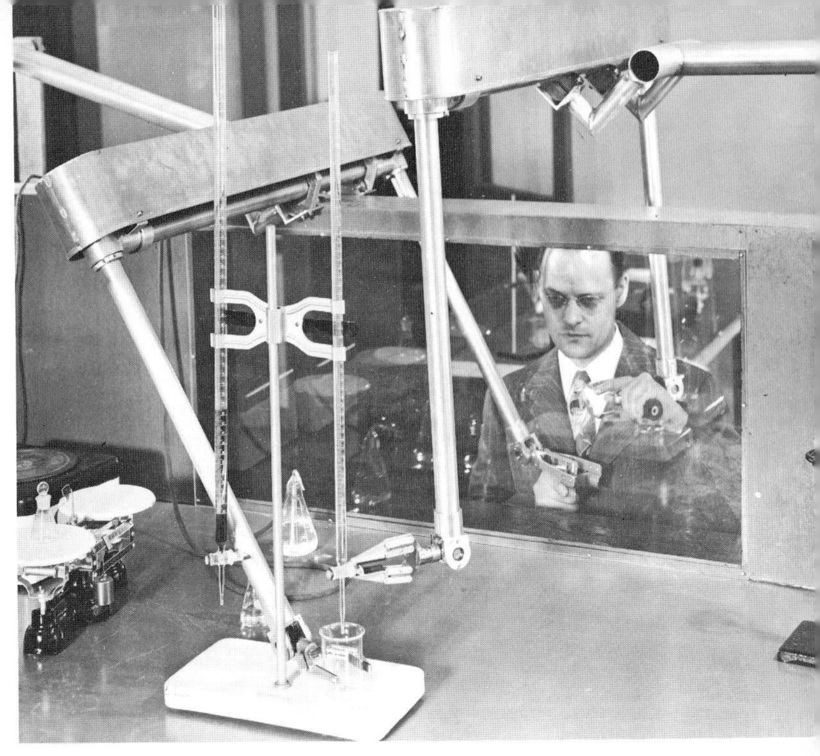

The first "master–slave" general purpose manipulator is being demonstrated by its inventor, Raymond C. Goertz, in 1949.

5-foot-thick concrete walls. This necessitated the development of new chemical-treatment methods, devised in the Chemical Engineering Division under Steve Lawroski, Milt Levenson, and their colleagues; the development of tools and techniques for making the fuel pins and putting them together into fuel assemblies, done in the Metallurgy Division under Frank Foote, Bob Macherey, and their colleagues; and the development of remote viewing and handling devices, done by the Remote Control Division under the late Ray Goertz and his colleagues.

Under the direction of Len Koch, project manager for EBR-II, Milt Levenson, Harry Monson, Wally Simmons, and their colleagues, the entire complex was built at NRTS in Idaho. Building such a complicated facility 1800 miles from home base posed problems quite aside from the technical ones, as those who were associated with the project well remember.

The reactor began operation in 1964, and the turbine generator was synchronized and first delivered power to the NRTS power loop on August 7. By the end of 1970, more than 250 million kilowatt hours of electricity had been produced by EBR-II. The AEC's and the nation's civilian power-reactor program was focused on the Liquid-Metal Cooled Fast Breeder Reactor (LMFBR), and the success of the LMFBR will rest heavily on the information obtained over the years from EBR-II.

The Chemical Engineering Division has in the past developed many methods for the processing of spent fuel from reactors—aqueous processes, the pyrometallurgical process, and the fluoride volatility process. Now, under the able direction of Dick Vogel, their attention has turned to the many chemical problems involved in using high-temperature sodium as a coolant in fast reactors. The Metallurgy Division in the past decade concentrated on development, and especially fabrication, of fuels for Argonne's reactors. Now, however, under the leadership of Paul Shewmon and Brian Frost and under a new name, Materials Science Division, they are concentrating on acquiring a very detailed knowledge of the behavior of fast reactor fuels and structural materials under the twin conditions of long-term irradiation and high temperatures. The Reactor Engineering Division, responsible for designing, engineering, and constructing so many of Argonne's reactors, has now been restructured into the Reactor Analysis and Safety Division and the Engineering and Technology Division. This reflects the concern with safety and the engineering development of components, which is such an important part of the LMFBR program.

Bob Laney was recently brought in as Associate Laboratory Director for Engineering Research and Development to assist in the refocusing and restructuring of Argonne's reactor program to reflect the nation's major reactor-development effort. His responsibilities will also involve the coordination of Argonne's increasing interaction with industry. I can assure you that in the decade ahead Argonne will continue to play an extremely important role in the AEC's reactor-development program.

In addition to its responsibilities in the reactor-development program, Argonne has from the beginning carried on a very fine and strong program of basic research.

Late in the 1950s the stage was finally set for a major effort that would widely expand opportunities for basic research in high-energy physics, not only for Argonne staff members but for high-energy physicists from Midwestern universities and from many parts of Europe. After 4 years of planning, ground was broken for Argonne's Zero Gradient Synchrotron (ZGS), a 12.5-GeV particle accelerator. On Dec. 4, 1963, it was my pleasure to participate in dedication ceremonies for this new tool, which was destined to contribute so much to the scientific life of the Midwest.

The ZGS was constructed in response to a long-standing need. Although large particle accelerators were available on the East and West Coasts, none was in existence in mid-America, and the high-energy physics departments of Midwestern universities were losing both faculty members and graduate students to institutions on the coasts.

The ZGS was designed to supplement, not to compete with, the machines already in existence here and abroad. Although its energy would not be as great as that of other accelerators, its intensity would be much greater. The ZGS incorporates a comparatively large aperture through which particles can pass. This "window frame" design combines with the high magnetic field to make possible the acceleration of large numbers of particles, providing a shotgun rather than a rifle approach to the creation of interactions that are of interest to the high-energy physicist.

Two important achievements resulted from the need for experimental apparatus to match the capabilities of the ZGS. One was the design, construction, and successful operation of the 12-foot bubble chamber, largest of its kind in the world, and another was the use of a superconducting magnet to power this huge chamber. Gale Pewitt presided over the birth of the 12-foot chamber, and John Purcell brought the big magnet into existence.

The value of the 12-foot chamber was demonstrated in November 1970, when, for the first time in history, a neutrino was observed in a hydrogen chamber.

Interior of 12-foot hydrogen bubble chamber, showing the tracks of particles produced by the neutrino interaction.

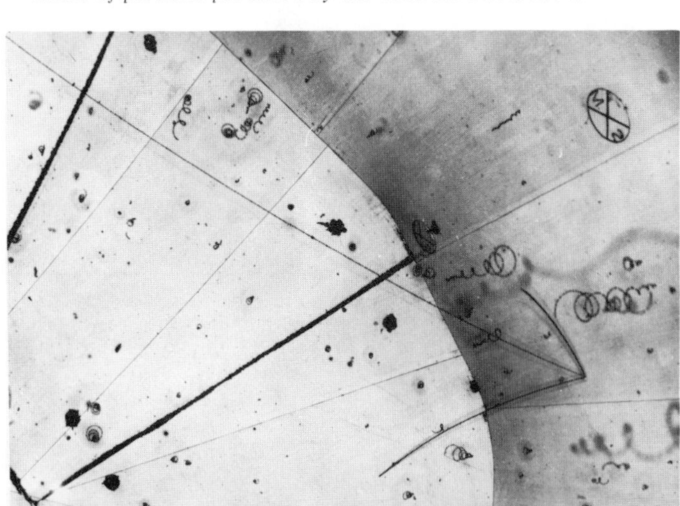

In the 6 years ending Dec. 31, 1970, 125 experiments had been carried out at the ZGS. Physicists from 50 universities had used the machine and had joined with Argonne staff members in the publication of 164 papers in professional journals.

The list of those who made important contributions to the development of the ZGS and the Argonne High Energy Physics Complex is a long one, and all cannot be noted. My early coworker and long-time friend Jack Livingood did the initial planning. Albert V. Crewe came aboard in 1958 to direct completion of the design and much of the construction. When Al became Laboratory Director in 1961, Lee Teng took over, and under his aegis the machine was completed. Ron Martin and the late John Fitzpatrick directed scientific and engineering activities; Martin Foss designed the magnet ring.

Through the decade of the sixties, the buck stopped at the desks of three Associate Laboratory Directors for High Energy Physics: Roger Hildebrand, Bob Sachs, and Bruce Cork.

During the sixties, under the leadership of Crewe and later of Robert Duffield, the results obtained in the areas of chemistry, physics, and materials research continued Argonne's reputation for high-quality research and added significantly to our fund of basic knowledge.

ANL directors (left to right), W. H. Zinn, Norman Hilberry, A. V. Crewe, and Robert Duffield.

During the sixties, under the leadership of Albert Crewe and later of Robert Duffield. . .

The Chemistry Division is an outgrowth of the Chemistry Section for which I had responsibility back in Met Lab days. Many of its present members were my wartime colleagues during my 4-year stay in Chicago. First under the directorship of Winston Manning, who was named Associate Laboratory Director for Basic Research in 1966, then under Max Matheson, and currently under Paul Fields, this division has been responsible for several important advances, among them:

- The discovery of the noble-gas compounds. In 1962 John Malm, Henry Selig, and Howard Claassen succeeded in combining xenon with fluorine to create xenon tetrafluoride, a relatively simple compound. The importance of this discovery derives from the fact that the noble gases had been thought to be inert and nonreactive.
- The discovery of the hydrated electron reported by Edwin Hart and his British colleague Jack Boag in 1963. The discovery and analyses of the roles of the hydrated electron and other short-lived fragments are leading to a better understanding of radiation chemistry.

Max Matheson, April 1965.

The Chemistry Division has been responsible for several important advances. . .

John G. Malm (left) and Howard H. Claassen prepare to remove a sample of xenon tetrafluoride from their experimental apparatus.

Crystals of xenon tetrafluoride.

Edwin J. Hart, research with the hydrated electron.

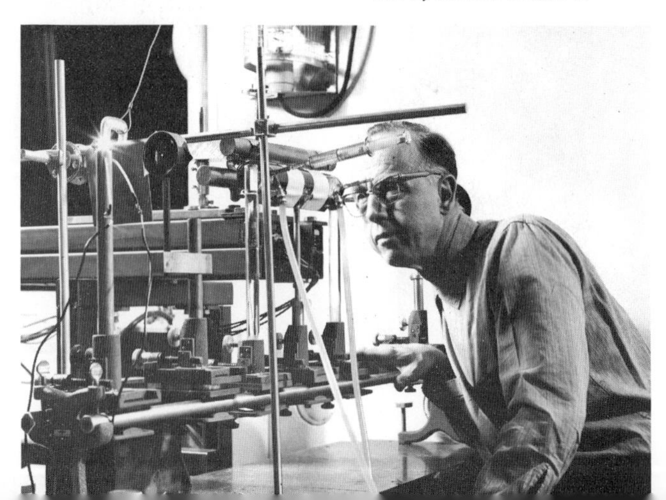

Joseph J. Katz explaining the use of simple living organisms, adapted to grow in essentially pure heavy water, to study life processes such as photosynthesis and protein metabolism.

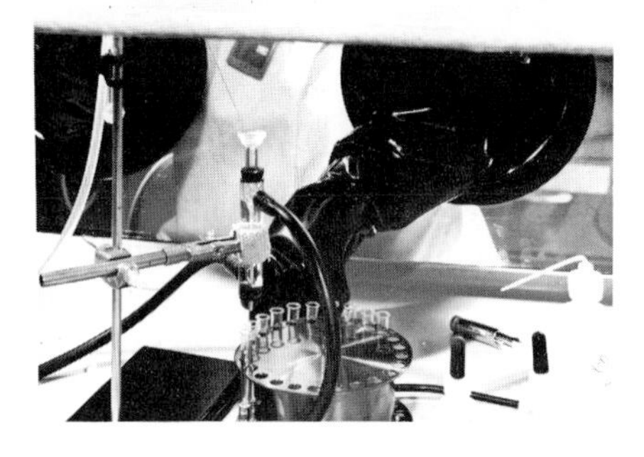

Paul R. Fields operates a small ion-exchange column in a glove box, a typical technique for separation of the actinides.

• Pioneering research in "isotopic substitution" in organic compounds, including the first complete substitutions of deuterium (heavy hydrogen) for ordinary hydrogen in living organisms, both plant and animal cells, by Joseph Katz and his group.

• Participation by Argonne chemists, notably Paul Fields and Martin Studier, in the discovery of some of the heavy transplutonium chemical elements. They also made unique contributions to the production, separation, and characterization of these elements and their isotopes.

Scientists are studying the Mössbauer effect in xenon tetrafluoride. The instrument is a mechanical velocity spectrometer.

Although in the past decade low-energy-physics research has been carried out under three different division directors, Lou Turner, Mort Hamermesh, and currently Lowell Bollinger, it has had the common thread of searching for a greater understanding of atomic structure. Among the first to initiate fundamental studies using the Mössbauer effect was Gil Perlow, who has built the technique into a powerful experimental tool in such diverse fields as nuclear structure, solid-state properties, and general relativity theory. Also, the angular-momentum-distribution discoveries of Schiffer and Lee have been of great importance in developing the field of nuclear spectroscopy; the discoveries of Erskine and others led to a better understanding of the nuclear properties of the actinides; and recent heavy-ion elastic-scattering studies are contributing significantly to nuclear-structure theory.

An understanding of the properties of materials has obviously been a strong interest of the atomic energy program, dating back to the Met Lab days, and it has become of increasing importance with the passage of time. Argonne has been, and continues to be, a leader in this field, having one of the largest combined basic and applied materials programs in the Western world. It started with the need to know the physical and chemical properties of fuels and structural materials under conditions encountered in reactors. Such work was initiated by personnel within the Metallurgy Division and the Chemistry Division. More recently the increased importance of a fundamental understanding of materials has been emphasized by Mike Nevitt, Paul Shewmon, and Norman Peterson and is also reflected in the recent renaming of the Metallurgy Division as the Materials Science Division. During the sixties the pure-research phase of this work finally came of age with the formation of the Solid State Science Division, which now occupies the newest of the major buildings constructed at Argonne. Under the direction of my Met Lab colleague Oliver Simpson, this work has taken on new importance.

The biological research program at Argonne is a natural extension of the biological work of the Met Lab. The potential danger of radiation was early recognized, and research into the biological effects of radiation on living organisms was among the earliest work started in the atomic energy program. The biological and medical research program at Argonne still has the same basic objective for which it was started.

The decade of the sixties has seen some changes, however. When in 1962 the Biological and Medical Research Division's director, Austin Brues, sometime artist, humorist, and world traveler but all-time biologist, expressed a desire to return to full-time research, his wishes were respected. He had carried administrative responsibilities since 1946. His successor was Max R. Zelle, a distinguished academician who, after 7 years as director, found a return to the university atmosphere irresistible. In early 1969 John F. Thomson, an 18-year veteran with the division, agreed to wear two hats until a candidate could be found. A little over a year ago, Warren Sinclair, a biophysicist, began a new era in the division's leadership.

Among the most important achievements of the past 10 years in the biological sciences have been comprehensive studies of the long- and short-term effects of a variety of types of radiation, on microbial, plant, and animal organisms. Attempts to modify radiation effects led to the development of the first successful protective agent against X rays, to the systematic exploration of chelating agents for removing radioactive metals from the body, and to basic studies in tissue transplantation and immunity mechanisms. Fundamental contributions have also been made in the study of aging and its relation to the late effects of radiation. These studies established the importance of the brain-to-body weight ratio as a determinant of species longevity. Current emphasis is on neutron effects studies with the Janus reactor, a facility capable of exposing large numbers of animals to neutrons without significant gamma-ray contamination.

The decade also saw a significant refocusing of the work of the Radiological Physics Division. John E. Rose was this division's director until 1963; then Leo Marinelli directed the division until 1967; and the present director is Robert Rowland. One of the earliest achievements of this division was the development of the first facility for pinpointing radiation in the human body with speed and accuracy. Argonne's "iron room" allows determination of the amounts, locations, and identities of extremely small quantities of radioactive materials in the body—as little as one billionth gram of radium. Similar facilities are now used throughout the world. Also of particular note has been its research on bone, both in the areas of bone physiology and of the effects of the radiation dose delivered by radioisotopes fixed in bone.

Early in Dr. Rowland's directorship the division embarked on a study of the sulfur dioxide content of the atmosphere over the city of Chicago. This was Argonne's first formal step in what has become a growing commitment to the solution of environmental problems.

In 1967 Congress broadened the Commission's charter to enable the AEC and its contractors to work with other agencies in the protection of public health and safety; thus Argonne has been able to undertake a broadened role as a major Midwest research center.

This has resulted in an accelerated interest in accepting new challenges, and in late 1969 the Argonne Center for Environmental Studies was established here. The center is designed to use an interdisciplinary approach to the achievement of three goals: first, to help gain a better understanding of the extent to which the environment is being changed; second, to define particular effects more quantitatively; and, third, to help with the formulation and presentation of varicus alternative courses of action.

This approach already has resulted in a model for predicting, analyzing, and controlling air pollution. Using studies of pollution emission from stationary sources as well as pollution dispersion patterns, Len Link and his colleagues developed a computerized model applicable to both the management of air-pollution emergencies and the long-range development of air-resource management. Their program presents guidelines for the creation of legislation, zoning ordinances, and tax incentives that would foster urban and regional growth in a manner compatible with acceptable air quality.

In 1968 Argonne began a study of heated discharges from power plants into large lakes. This program is establishing a mathematical model of circulation patterns in Lake Michigan, developing models to express the behavior of thermal plumes, and analyzing the mass–energy balance of the lake. The study also outlines the research needed for the understanding of thermal effects on the ecosystem so that methodologies can be provided. This work is expected to have a strong bearing on reactor–siting criteria.

Two other Argonne programs are of special interest: One is the development, by the Laboratory's Chemical Engineering Division, of fluid-bed techniques in the combustion of coal. Use of these techniques could reduce emission of sulphur dioxide into the atmosphere. The second is work on lithium anode secondary cells, also being carried out by the Chemical Engineering Division. Such cells promise to be useful as a primary source of power for automobiles and to have dramatic possibilities as an implantable energy source for individuals with heart defects.

The change in the AEC's charter also made possible "spin-off" activities that give great promise of providing benefits for mankind. These include:

- A hemodialyzer (artificial kidney), developed by Finley Markley of the High Energy Facilities Division and A. R. Lavender of Hines Hospital, which may revolutionize the care of patients suffering from kidney disease. Victims of kidney failure now must depend on very complex and expensive hemodialyzers that can be used only at hospitals. The new kidney machine is so inexpensive, small, and simple that it may be possible for the patient to use it himself at home. The device was made possible through the use of adhesives Markley developed for application in the construction of the ZGS.
- A Braille machine, developed by Arnold Grunwald of the

Finley Markley, with the artificial kidney he developed.

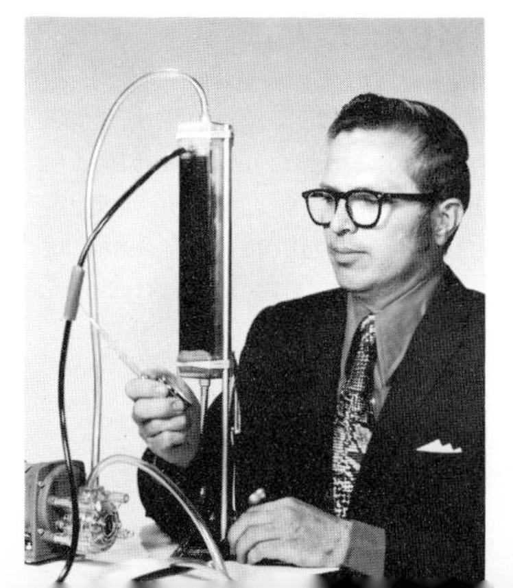

Arnold Grunwald, with the Braille Reader he developed.

Engineering and Technology Division. Smaller than a portable typewriter, it will take symbols recorded on ordinary magnetic tape and play them back on an endless plastic belt in raised dots forming letters in the Braille alphabet. It will reduce by a factor of 250 to 500 the bulk of Braille materials to be produced, handled, and stored, permitting much wider use of Braille literature by the sightless. This development is being supported under a grant by the U. S. Office of Education.

When Argonne was established as a national laboratory, the Commission and the Argonne administration agreed that interaction with the academic community would be a primary responsibility of the laboratory.

In 1950 Joe Boyce attacked the problem, and the foundation he established in the following 5 years made possible a program that flourished in the decade of the sixties.

The initial organization through which the laboratory sought to interact with universities and colleges was the Participating Institutions Committee, organized very early in Argonne's history. Thirty-two Midwestern universities were members. Through several intermediate steps, this organization evolved into Associated Midwestern Universities, Inc. (AMU), incorporating in its membership 30 universities.

At this time Frank Myers gave up his post as Dean of the Graduate School at Lehigh University to become Argonne's Associate Director for Education. Shortly afterward John Roberson took over as Executive Director of AMU. These events resulted in new impetus to educational activities that brought Argonne and the academic community into closer association.

Still another change occurred in 1966—one that would give universities an even stronger role in the activities at Argonne. In that year Argonne Universities Association (AUA) came into existence, and a new 5-year contract for the management of Argonne stipulated that AUA, the University of Chicago, and the Commission would share in management responsibilities.

Under the terms of the contract, AUA formulates, approves, and reviews laboratory programs and policies. The University of Chicago, which has operated Argonne from the time it was founded in 1946, continues to be responsible for its management and operation in accordance with the policies established by AUA. The Commission, of course, has provided a major share of the laboratory's financial support and participates in major decisions affecting Argonne's welfare.

Thirty universities now hold membership in AUA.

The most recent change in the mechanism for fostering Argonne–university interaction occurred in 1968. In that year all of Argonne's educational activities were placed under the direction of a Center for Educational Affairs, and Shelby Miller came to Argonne from the University of Rochester to become Associate Laboratory Director for Educational Affairs and Director of the Center.

Progress in this area has been so rapid that the Center was able to report that last year 2600 university and college representatives—college juniors up through faculty members—participated in activities at Argonne.

The record would not be complete without my recalling one of the most dramatic ventures in education this nation has ever undertaken. In 1953 President Eisenhower used the vehicle of his famous "Atoms for Peace" talk to suggest that this country establish means for sharing with many nations of the world our rapidly growing understanding of the peaceful uses of nuclear energy. Argonne considered this a mandate and launched a crash program to bring into existence the International School of Nuclear Science and Engineering. Norm Hilberry, Elmer Rylander, and Rollin Taecker did yeoman work, and, before the year was out, the school was in operation. Its objective was to attract young men from abroad and to provide them with sufficient training to enable them to return home and establish nuclear energy programs appropriate to the level of technology existing there.

In 1961 the International School became the International Institute. In the institute the emphasis was on programs tailored for each participant to make maximum use of the background and the skills he had already acquired. It was the continued success of the institute which caused its demise in 1965. So many of its graduates scattered about the globe had developed strong nuclear energy programs in their home countries that the kind of training offered at Argonne was no longer needed.

As most of you here today will recall, Al Crewe decided to step down from his position as laboratory director in December of 1966, and early in 1967 Robert Duffield, whom I have known since our association during his student days at the University of California at Berkeley, succeeded him as director. Bob Duffield has continued the fine tradition of leadership here at Argonne, guiding ANL through a significant and productive era of its history.

Let me emphasize that the projections the Commission has

A major source of long-term national strength. . .

developed indicate an undiminished need for use of Argonne National Laboratory for AEC programs for as far ahead as we can make projections. I foresee no lessening in the national importance of the sort of work Argonne has been carrying out for the AEC. I understand that, in addition to the support we provide, the support for work at Argonne funded by other agencies will total about $2,500,000 this fiscal year. The Commission will continue to encourage its laboratories to provide assistance to others in areas in which they have special competence and facilities up to the limits set by statute and the priority we need to give our own work.

Argonne will continue to play a central role in what I see as perhaps the most fruitful and in many ways the most exciting technological challenge facing the nation today—the development of the breeder reactor. Further, I believe that pioneering research at Argonne in both the physical and biological sciences will gain continued recognition as a major source of long-term national strength.

The drive for excellence in any laboratory is fueled most simply by rapidly expanding requirements and budgets. For now we must find a way to maintain our momentum with different fuel. This is a time of testing for many scientific institutions. Some will be seized by the mincing caution that chokes inventiveness. Some will wander and wither, seeking the favors of fashionability instead of capitalizing on their own virtues. Certainly the future of Argonne will be affected by decisions made elsewhere and by the priorities others attach next year and the year after to specific efforts. For the long run, however, I view decisions by individuals here about their own work as of even greater importance. The best assurances for the future will come from present rededication to the drive for excellence that Argonne National Laboratory has displayed throughout its first 25 years. ■

THE SAVANNAH RIVER PLANT

At the celebration, Curtis A. Nelson, first AEC manager at the site; Glenn T. Seaborg; and J. A. Monier, the present plant manager.

A Twentieth Anniversary

The Twentieth Anniversary Observance of the Savannah River Plant Sponsored by the South Carolina–Georgia Nuclear Council, Augusta, Georgia, Dec. 7, 1970

■ It is a distinct honor and a pleasure to join this gathering in celebration of the twentieth anniversary of the Atomic Energy Commission's Savannah River Plant.

As its very able manager, Nat Stetson carries a heavy responsibility. The plant today represents an investment of some 1.3 billion dollars and has a total of about 5800 federal and contractor employees and an annual budget of 116 million dollars. I need not add that this facility is an important part of the economy of Georgia and South Carolina and an indispensable part of AEC operations.

When one considers anniversaries, it is usually in the context of marriage, and whoever originated the symbols for such yearly celebrations devised an interesting list of materials for gifts that increase in value as one gets older. We find that the proper gift for a couple married one year should be made of paper; for 10 years, tin; and for 20 years, china. Savannah River is producing a material on its own which outranks in value all the gifts on the list, even the diamonds appropriate to a seventy-fifth anniversary.

I refer, of course, to californium-252, a radioisotope that holds great promise. When we first announced the availability of this rare isotope two years ago, the price was 450 billion dollars a pound—if a pound had been available. Fortunately, in californium we have, in Christopher Marlowe's phrase, "infinite riches in a little room," and only microscopic amounts are needed. My point is that any gift we might present on this anniversary would seem modest indeed beside the value of the product the Savannah River Plant contributes to the nation.

Perhaps, as we look back, 20 years seems insignificant as a period of time. What counts, however, are the events occurring in any given era. And the past two decades have certainly been packed with significant change.

To appreciate that era and this anniversary, we must go back briefly to 1950. Where were you, and what was going on 20 years ago when the decision was made to build the Savannah River facility? Harry S. Truman was President. Hopalong Cassidy was the top star on television. Fifty-cent pieces were plentiful. The Dodgers were in Brooklyn, and the Giants were in Manhattan. The Atlanta Falcons did not exist, and, even if they had, you could not have seen them on color televison because it was not around either. In 1950 the Supreme Court said in a historic ruling that under the Fifth "infinite riches in a little room. . ."

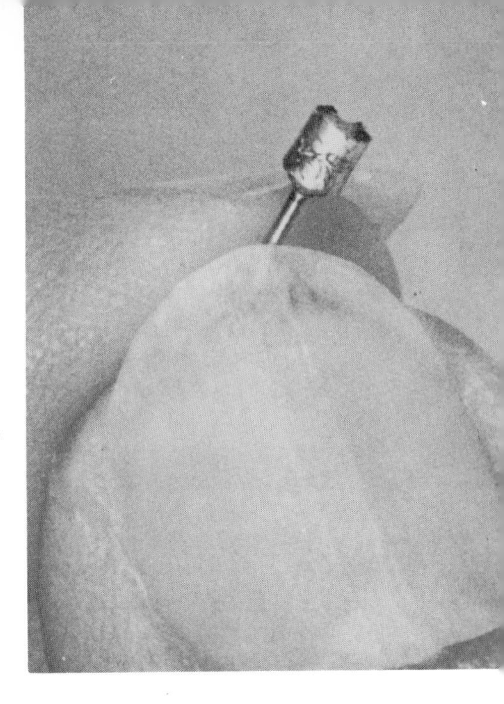

A pellet of compacted platinum foil of the size used to contain a microgram of californium-252. (The actual source would be too radioactively "hot" to handle.)

The first pure californium compound to be isolated in the laboratory. Consisting of three ten-millionths of a gram of californium oxychloride, the compound was isolated and identified by Burris B. Cunningham and James C. Wallmann in the University of California's Lawrence Radiation Laboratory. The sample, magnified about 170 times, is contained within a glass tube and gives off its own light as a result of radioactive decay.

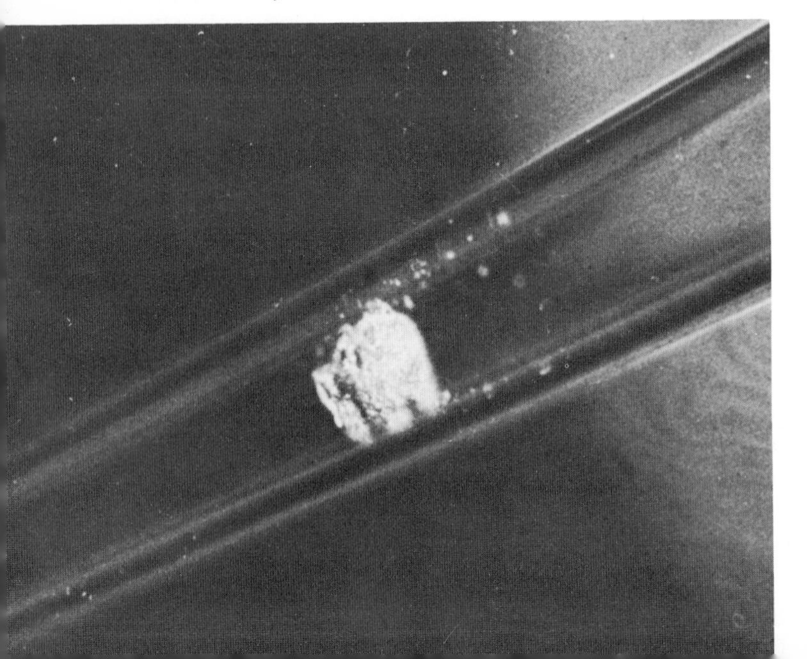

Amendment no one could be forced to testify against himself. On Nov. 28, 1950, the U. S. Atomic Energy Commission took a major step with the announcement of a site in South Carolina to build a facility that over the course of two decades not only contributed substantially to the defense of the United States but also produced nuclear materials beneficial in industry, medicine, and agriculture.

To obtain the full flavor of the events and decisions leading to the Nov. 28, 1950, announcement, we must go back to Sept. 3, 1949. The events of that day and the subsequent developments have much drama.

"It is not given to human beings," Winston Churchill once wrote, "to foresee or predict to any large extent the unfolding course of events. In one phase, man seems to have been right, in another he seems to have been wrong. Then again, a few years later, when the perspective of time has lengthened, all stands in a different setting, there is a new proportion. There is another scale of values."

On Saturday, Sept. 3, 1949, a new scale of values began developing for the United States. In Washington on that day, the U. S. Senate set a new record for short sessions when it succeeded in assembling and adjourning in forty seconds. Like others, the senators were anxious to leave the Capitol for the Labor Day weekend. By late afternoon most of the central city was deserted. Even the traffic on Pennsylvania Avenue in front of the White House had subsided to a few automobiles. On G Street, just west of the Executive Mansion, the office buildings were empty except for a few guards and an unlucky group of Air Force officers and enlisted men who had drawn duty on the last holiday of the summer. As the slanting rays of the afternoon sun pierced the clouds, the staccato rhythm of a teletype machine broke the drowsiness. No one suspected that the report sputtering from the machine would set in motion a chain of events placing on the Atomic Energy Commission and other segments of the U. S. Government a burden of extraordinary decisions. The tangle of events of the next five months recorded more than a political struggle; they seemed to involve the very destiny of man.

The teletype report alerted the headquarters of the Air Force Long Range Detection System that a WB-29 weather reconnaissance plane on routine patrol from Japan to Alaska had picked up some measurable radioactivity.

By Monday morning, September 5, there was enough additional information to spoil the holiday for most of the Long Range Detection staff. Reports of substantially higher counts of radioactivity began coming in. The British checked the radioactive air mass as it proceeded east. Officials were soon convinced that the Soviet Union had indeed detonated a nuclear device.

The event was determined to be a successful test explosion by the Soviet Union of its first atomic bomb. The news was announced to the American public by President Truman on September 23.

The Russian explosion, narrowing as it did the superiority of the United States in the atomic field, prompted President Truman as he announced the confirmation to remark privately that "this means we have no time left."

In Washington, on Nov. 9, 1949, when the first formal official step was taken to bring order out of the chaos of dispute, the day was cloudless and pleasant.

At 5 p.m. that afternoon, a top secret document was dispatched to the White House by the Atomic Energy Commission. After reviewing the recommendations, President Truman on November 19 directed a letter to the Secretary of the National Security Council appointing a committee composed of the Secretary of State, the Secretary of Defense, and the Chairman of the Atomic Energy Commission to study further the Commission's conclusions and to suggest steps that should be taken by the White House.

Within several weeks the Committee voted to recommend that the United States proceed with a nuclear effort. The ensuing months brought an intensive study by the Atomic Energy Commission of the problems involved in construction of a plant to produce materials needed for a thermonuclear device.

On June 12, 1950, the Commission approached the E. I. du Pont de Nemours Company, which had built the plutonium factories in the State of Washington during World War II, to consider a contract for the construction of reactors of a new design to produce the necessary materials. Within a few days Du Pont had in its possession a letter from the President urging participation in the interest of national security. On August 1, Du Pont signed a contract with the Commission to proceed with the design, construction, and operation of nuclear reactors that could produce materials important to the nation's defense.

Meanwhile, the Commission had requested that the Corps of Engineers, United States Army, aid in a survey of a new production site and in the acquisition of land once a site was chosen. On June 29, the Commission, Du Pont, and the U. S. Corps of Engineers formally began the site survey.

When the Commission announced on Aug. 2, 1950, that Du Pont

had agreed to a contract to design, construct, and operate new atomic facilities, it stated the work would be done at a "site yet to be selected." This statement brought to the offices of the Commission between that date and October 15 more than 1100 letters from interested Americans with recommendations for 147 different sites.

A scene in the quiet little farming town of Dunbarton, South Carolina, several months after the announcement that Dunbarton and other small towns in the area must be evacuated to make room for the AEC's giant Savannah River Plant.

The quiet tree-lined main street in Ellenton, South Carolina, at about the same time.

Du Pont, the Commission, and the Corps of Engineers reviewed the merits of 114 potential sites in nearly every section of the country. The list was eventually pared to seven, and in early November a final decision was made.

On November 17, Gordon Dean, then Chairman of the Commission, went to the White House to advise the President of the decision. The President, looking at a map of the proposed site of more than 200,000 acres that was to be taken over, asked whether a less populated area might not be found and instinctively pointed out the hardships to be faced by communities such as Ellenton and Dunbarton.

A reluctant "No" came from Mr. Dean. The site best met the criteria set up by the Commission and the National Defense Establishment. The criteria for such a site had been carefully worked out by the Commission and Du Pont. First, there was need for great amounts of water to cool the nuclear reactors that were to be built. The Savannah River, one of America's historic rivers, the river DeSoto first saw in 1540 when he crossed it just south of Augusta in a search for gold, was considered an ideal source. Other factors considered were a low-population area near high-population centers, freedom from floods, a satisfactory power supply, accessibility, transportation, terrain, low incidence of storms, the operating requirements of the plants themselves, and the fact that construction was possible the year round.

November 28, 1950, was unseasonably cold in the central Savannah River area. The temperatures had plummeted to below freezing, and the weather was in sharp contrast to the news that was in the making. On the day before, Sunday, a group of Atomic Energy Commission and Du Pont officials from Oak Ridge, Washington, and Wilmington had met in Augusta to put the final touches on plans to announce the beginning of a new era and then had split up into teams to give the news to officials of cities and communities in the immediate area on the next day. Some persons in the central Savannah River area might have known something was in the air, but they were not prepared for this news at 11 a.m. on November 28:

> The United States Atomic Energy Commission today announced that its new production plants to be designed, built, and operated by E. I. du Pont de Nemours & Co., Inc., of Wilmington, Del., will be located in Aiken and Barnwell counties, South Carolina. The new site will be known as the Savannah River Plant.
>
> To make way for the plants and the surrounding security and safety

The new site will be known as the Savannah River Plant. . .

---

zones, it will be necessary for some 1500 families to be relocated in the next 18 months . . . .

Although 20 years have passed and our country has performed great feats in its space program by sending men to the moon, the magnitude of the job of building the Savannah River Plant is still

The Savannah River, which gives the plant its name, follows the border of the property for 22 miles.

1500 families had to break long ties
and move to new locations. . .

Curtis A. Nelson, then manager of the AEC's Savannah River Operations Office (left), and Robert K. Mason, field project manager, Du Pont Company, in late November 1950, talk with Ellenton's mayor, H. W. Risher, about the evacuation of his town.

This crudely written message was nailed up during the night after the announcement—Nov. 28, 1950—that Ellenton was in the area to be evacuated to make room for the Savannah River Plant.

First home to be moved out of the area in 1951 belonged to Hampton Irwin. Here the Irwin family look at the strange sight of their house jacked up on a trailer just before it was moved to a new location 16 miles off the site. Owners were reimbursed for their property.

AEC and Du Pont Company officials explain to residents of the Savannah River Plant area details of how their property is to be taken over, early December 1950.

"The Long Store," as Cassells Company, Inc., was called, a creaky wooden structure 210 feet long and 41 feet wide, was probably the champion of all country stores before Ellenton, South Carolina, was evacuated to make way for the Savannah River Plant.

impressive. Not so well known are several interesting sidelights which have become a part of the history of the Savannah River Plant.

How many, I wonder, know of the contributions made during construction by the David brothers—two 4-foot 2-inch midgets whose previous claim to fame had been their appearance in the motion picture *The Wizard of Oz*. They were employed in the Savannah River Projects Pipe Department as welders since they were able to work in cramped quarters that were inaccessible to men of normal size.

Then there was the son of Vachel Lindsay, the famous American poet, who put in a stint as a carpenter, and Jim Nabors, television's Gomer Pyle, who spent a short time at the plant as a clerk in the early 1950s. In addition, the Savannah River Plant was a lifesaver for a number cf sponge divers from Tarpon Springs, Fla. These men, who had run into hard times because of a falling market in sponges, were brought to the Savannah River Plant by a Du Pont subcontractor as painters.

We all appreciate, of course, the immensity of the project that brought nearly 39,000 workers to your area at the peak of construction in September 1952, but only a summary can adequately reflect the tremendous extent of the work. The statistics remain impressive. The labor and materials included:

39,150,000 cubic yards of earth moved; equal to a wall 10 feet high and 6 feet wide from Atlanta, Ga., to Portland, Ore.

1,453,000 cubic yards of concrete poured; equal to a highway 6 inches thick and 20 feet wide from Atlanta, Ga., to Philadelphia, Pa.

118,000 tons of reinforcing steel; equal to 3300 carloads or a train 30 miles long.

27,000 tons of structural steel; equal to a train 8 miles long.

85,000,000 board feet of lumber; enough for a city of 15,000 homes with an average population of 45,000.

126,000 carloads of other materials; equal to a continuous train from Atlanta, Ga., to New York City.

52 miles of water pipeline about 84 inches in diameter.

230 miles of new roads and 63 miles of permanent new railroad track built.

2,000,000 blueprints; equal to a strip of paper 24 inches wide reaching from Atlanta, Ga., to Seattle, Wash.

For assistance in design and construction, Du Pont retained subcontractors, which included, among other industrial leaders, such well-known firms as Blaw–Knox Company, the Lummus Company, American Machine & Foundry, New York Shipbuilding, Combustion Engineering, B. F. Shaw Company, Johns–Manville, and Miller–Dunn Electric.

The translation of these materials into the complex structures and supporting facilities which make up the Savannah River Plant brought traumatic experiences to many citizens—the 1500 families that had to break long ties and move to new locations, the residents in the surrounding area who had to cope with the influx of thousands of new persons and to adjust to changing conditions.

It is a matter of record, however—and this is borne out by Atomic Energy Commission and Du Pont officials familiar with such large construction projects—that the cooperation, the dedication to national purpose, and the positive attitudes of the residents of this area during construction and operations were and are most outstanding. The public support of the construction and operations of the Savannah River Plant and the citizens' interest in plant developments, including those involving peaceful uses of the atom, have been among the most salutary and rewarding ever experienced by the Atomic Energy Commission. What has impressed the Commission is the confidence shown by the public in the integrity of the personnel who operate the plant and in the positive programs that are maintained to protect the health of the environment. There are indeed environmental considerations in the control and containment of radiation, and the public in this area appreciates that every effort

This was a main intersection in the little town of Ellenton shortly after the area was taken over for the Atomic Energy Commission's Savannah River Plant.

The David brothers—two 4-foot 2-inch midgets—were employed as welders since they were able to work in cramped quarters that were inaccessible to men of normal size.

The project brought nearly 39,000 workers to the area at the peak of construction in September 1952. . .

"The Long Store" became a lunch stop for workers during early construction days of the Savannah River Plant.

South Carolina statesman James F. Byrnes (center right) dons a hard hat to tour the Savannah River Plant during construction days. Accompanying him are (left to right) Robert K. Mason, Du Pont field project manager for the Savannah River Plant; Monte Evans of the Du Pont Company; Donald A. Miller, former Du Pont manager of the Savannah River Plant; and (far right) Curtis A. Nelson, then manager of the Atomic Energy Commission's Savannah River Operations Office.

Badge check is first on the agenda for Du Pont Company President Crawford H. Greenewalt on his arrival on Sept. 26, 1952, at the Savannah River Plant. He is followed by General Manager H. F. Brown and Assistant General Manager R. M. Evans.

is made to operate the plant safely and constructively and that meticulous, continuing attention is given to such programs. On behalf of the Atomic Energy Commission, I would like to thank you for your support, cooperation, and understanding.

The competence and dedication of the Du Pont Company in the design, construction, and operation of the Savannah River Plant have indeed been noteworthy. The company has assigned some of its most capable administrators and technical personnel to plant programs.

For example, former President Crawford H. Greenwalt, now chairman emeritus of the board of directors, was, from the beginning, personally involved in studies and planning for the project. Assignments in the organization set up by the company to oversee operations of the facilities included such outstanding men as R. Monte Evans, William H. Mackey, and Lombard Squires, all of

---

The Du Pont Company President, Crawford H. Greenewalt, was, from the beginning, personally involved in studies and planning for the project. . .

whom spent considerable time on the site in meeting construction and operations schedules. The travel they and their associates logged on round trips from Wilmington to the plant site totaled thousands of miles and thousands of hours.

In operating one of the world's most complex facilities, Du Pont has efficiently discharged its responsibilities. The company has consistently met or exceeded schedules for delivery of nuclear products for national defense, continuously worked on and brought about improvements in plant processes and operations, generated economies that have resulted in savings of the taxpayers' dollars, and worked assiduously to develop projects to realize most effectively the potential of the atom for peaceful purposes and to maintain the health and vitality of operations.

In two other areas of interest, the company has achieved safety records that are nationally and internationally recognized and has encouraged its employees to participate in every aspect of community activity and development. The staffs at the Savannah River Plant, both Du Pont and the Atomic Energy Commission, are to be commended for the leadership and interest they have shown in community progress and uplift.

Heavy-water production at the Savannah River Plant.

Among those who have set the tone for community relations and participation in civic activities are Don A. Miller, first plant manager, and Julian Ellett and Jay Monier, who, in turn, succeeded him, all of Du Pont; Curtis A. Nelson, first AEC manager at the site, and Robert C. Blair and Nat Stetson, who followed him; and Gus O. Robinson and Howard Kilburn of the AEC. They have been joined by Milton H. Wahl, first director of the Savannah River Laboratory, and his successors, Bill Overbeck, Frank Kruesi, and Clark Ice, and by Robert K. Mason, first construction manager at the site, and his successors.

In addition, they have endeavored at all times to keep the public promptly and adequately informed. As a result, the public has confidence and faith in the integrity of management and in the programs that are carried on.

The record of our contractor, the Du Pont Company, and of the Commission's Savannah River Operations Office in effectively pursuing new ways to utilize the large investment at Savannah River constitutes an outstanding chapter in the story of nuclear energy, and we expect that the coming years will bring new and useful achievements for the American public.

The coming of the Savannah River Plant to this area set the base for acceleration of industrial development that has made the central Savannah River area one of the nation's most dynamic sections, reflecting outstanding cooperation between the residents of two states, who take advantage of a river as a state line rather than allowing it to isolate their interests. An example of such cooperation is, of course, the formation of the South Carolina–Georgia Nuclear Council, organized in 1967, which has as its objectives "maintenance, on behalf of the counties and communities within the vicinity of the Savannah River Plant, of the furtherance of continued economic progress of South Carolina and Georgia" and "serving as a focal point for expressing the interest and cooperation of the various communities in the development of peacetime atomic energy projects both in and outside the confines of the Savannah River Plant." The Council has effectively helped to create in the public's mind an appreciation of the constructive programs of the plant.

During the 20 years since the announcement of the site for the Savannah River Plant, the U. S. Government has contributed approximately three billion dollars to the economy of this area through plant construction and operations. New industries, such as Owens–Corning Fiberglas, Pyle-National, Kimberly–Clark, Olin Industries, Columbia Nitrogen, Proctor and Gamble, Continental

Can, American Cryogenics, Du Pont, and others, have created new jobs and new opportunities.

The most recent addition to the growing list of industries is Allied-Gulf Nuclear Services, which will construct a 70-million-dollar nuclear-fuel-reprocessing facility in Barnwell County adjacent to the Savannah River Plant. In connection with this development, there is no doubt that the positive attitude of residents of this area on nuclear programs was one of the compelling factors leading to the selection of the site. Furthermore, with Barnwell County taking title from the U. S. Government to an additional 900 acres bordering the 1587-acre Allied-Gulf site, the way is open to attract other nuclear-related activity. I understand that the Council is planning to further develop the concept of a nuclear center. The future could be most rewarding for the entire central Savannah River area in both South Carolina and Georgia.

Mankind's growing concern for his environment has opened new avenues of usefulness for the Savannah River Plant with its vast acreage of woods and waters. The Commission is pleased that the site has become the locale for ecological studies that will undoubtedly uncover useful information for both nuclear and nonnuclear industries. These studies are being pursued at the Savannah River Ecology Laboratory, operated for the Commission by the University of Georgia. Du Pont also is active in this area at the plant.

You are aware, of course, that the Savannah River Plant is still involved in the production of materials for the national defense. What is heartening for the future, however, is the continued utilization of the versatile nuclear reactors here for the production of materials that will directly benefit humanity. The Savannah River Plant is destined to play a continuing significant role in the constructive development of nuclear energy.

No history of the Savannah River Plant would be complete without mention of a major step forward in 1965 when one of its reactors reached a sustained neutron flux of record proportions. This flux totaled 6.1 quadrillion neutrons per square centimeter per second. To give you an idea of the achievement, this was more than 100 times the flux in most nuclear power reactors.

The significance of this accomplishment was that it opened the way for production of isotopes of very high specific activity, for production of man-made elements heavier than uranium which have important research potential, and for the speed-up in production of special radioactive isotopes.

The potential for good inherent in such materials as californium-

An enviable record in responding to national needs. . .

252, plutonium-238, cobalt-60, and curium-244 is tremendous. With capabilities not found elsewhere, the Savannah River Plant has played a unique role in nuclear development through its production programs. The plant produced the first commercial quantities of plutonium-238 and curium-244 and the first significant quantities of californium-252 and is the principal source of high-intensity cobalt-60. The plant has set an enviable record in responding to national needs. In the future its reactors can be expected to meet Commission requirements with the same degree of competence and with the same dedication to safety that has marked past and present operations.

At the beginning of my remarks, I mentioned the great monetary value of a small amount of the heavy-element radioisotope californium-252 produced at the Savannah River Plant. The true value of this isotope is yet to be appreciated in terms of its uses to mankind. The uniqueness of californium-252 as an investigative and research tool is expected to bring a variety of beneficial applications, such as improving health, upgrading industrial products, increasing energy reserves, and expanding the nation's supply of valuable metals. Californium-252 can be used in cancer research and treatment, industry, space exploration, general scientific research, civil engineering, agriculture, mineral exploration, and hydrology and petroleum exploration.

It is apparent that the unique properties of californium-252 already have been widely recognized. Since the AEC announced its market-evaluation program for this isotope in June 1969, the Savannah River Operations Office has received more than 2000 requests for information on its potential uses.

Californium-252 has two characteristics that make it especially valuable. First, it lasts a reasonably long time compared with other isotopes. It has a half-life of 2.6 years. That means we can ship it to hospitals and industries with the assurance that it will not quickly fade away before it can be put to work.

Second, this isotope emits a prodigious quantity of neutrons, the submicroscopic "bullets" that cause the splitting of uranium atoms in atomic furnaces. A tiny quantity of californium, just 1 curie of it, weighing about 2 milligrams, emits almost as many neutrons per square centimeter per second as a reactor. It is for this reason that californium has been called a "hip-pocket reactor."

The advantages of this portability are immediately apparent. With such a compact and lightweight source, hospitals no longer need to send cancer patients to an accelerator or reactor for neutron irradiation but can treat them in their own buildings. Other users will find its diminutive size equally advantageous.

The list of applications for this new product continues to grow. The U. S. Department of Agriculture Sedimentation Laboratory is planning to evaluate the potential of californium-252 in measuring moisture and density of soils and for chemical analysis of certain elements in the soils. At Kansas State University, californium may help cope with a new environmental problem. The material is being evaluated there for quantitative analysis of mercury and other toxic materials in commercially processed food.

Several hospitals in the United States are testing californium-252 for use in the treatment of cancer; the material is also being used in well logging to locate oil and gas reserves and in programs to locate gold and silver underground and minerals on the ocean floor. In another area, californium-252 is being used in a metering device to measure the sulfur content of coal as it moves on a conveyor belt. The objective is to reduce air pollution, since sulfur is a prime contaminating element. If sulfur in coal can be reduced or eliminated, cleaner air and better health will result.

Sources of californium-252 fabricated by the Savannah River Laboratory are now being loaned to 26 commercial firms and private institutions. In addition to the loan program, californium-252 has now been placed on sale more broadly, since larger quantities will be available at a reduced price in early 1971 through a special production program in one of the Savannah River Plant reactors. The larger quantities I mention consist of a total supply of about 1 gram—or $\frac{1}{28}$ of an ounce.

Another important isotope made at the Savannah River Plant is plutonium-238. Plutonium-238 holds great promise as an energy source for a nuclear-powered cardiac pacemaker. Such a pacemaker, using material produced at Savannah River, was successfully implanted in a dog in 1969 at the National Heart Institute at Bethesda, Md. Scientists hope that the nuclear-powered pacemaker will eventually replace the battery-operated devices currently used by about 20,000 persons, and that it will be used by the approximately

All these are products of the Savannah River Plant. . .

5000 new patients incapacitated by heart block each year. The nuclear pacemaker is about two-thirds the size of a pack of cigarettes and weighs about three and one-half ounces. Its longer life gives it a great advantage over the battery-operated pacemaker, which must have its implanted batteries replaced by a surgical procedure every few years. Nuclear-powered pacemakers have already been implanted in human patients in France and England.

The Savannah River Plant also has contributed to research on a nuclear-powered artificial heart. In cooperation with the Los Alamos Scientific Laboratory, plutonium-238 prepared in one of the Savannah River reactors has been refined and encapsulated as a heat source for such a device. If it is successful, the program would make a significant contribution to extending the life of heart patients; physicians have estimated that more than 10,000 artificial hearts could be used annually in the United States alone.

The same radioisotope that would power an artificial heart is already supplying power in space. Both Apollo 11 and Apollo 12 astronauts left several scientific devices on the moon which use plutonium-238 as a heat and power source. Through the services of such devices, scientists are learning more about the history and makeup of the moon. The material is also being used as a source of electric power in weather satellites that will enable scientists to give us more accurate long-range weather forecasts.

Two other useful products produced at the Savannah River Plant, cobalt-60 and curium-244, have potential for space heating in remote frigid locations as well as underseas habitats and as sources of electric or mechanical power in space, terrestrial, and underseas environments.

All these are products of the Savannah River Plant, and all are products with a future. We have seen only the beginning of the age in which their many uses will be applied and their benefits to mankind realized.

So the 20 years of this great nuclear-age facility whose anniversary we now celebrate have been only a prologue to a fascinating story yet to unfold. We can be proud of what we have built here. And we are grateful to all of you who have helped build and operate this important plant. Now we must focus on the future with the same determination and dedication that has brought us this far. It will not be a future without its share of obstacles and its demand for hard work, perseverance, imagination, and innovation. But the Savannah River Plant provides a good background and a good base on which to build that future. I hope that 20 years hence we can again gather to celebrate an anniversary of this plant and at that time can look back on an even greater era of progress than we are privileged to celebrate today. ■

HFIR, TRU, *and* TRL

A Transuranium Team

Dedication of the Transuranium Research Facilities, Oak Ridge National Laboratory, Oak Ridge, Tennessee, Nov. 8, 1966

This is a great occasion for nuclear science and for many nuclear scientists—this one in particular. I am especially pleased—perhaps three times as pleased—that you have invited me to participate in this triple dedication. It is not often that I get the opportunity to dedicate three facilities with one talk. I am also happy that a number of honor students from the high schools and colleges of the area and their teachers are present this afternoon.

The fact that we are dedicating the High Flux Isotope Reactor (HFIR), the Transuranium Processing Plant (TRU), and the Transuranium Research Laboratory (TRL) is, of course, of personal significance to me. As you know, I have a special attachment for the transuranium elements. I was going to say "fondness" rather than "attachment," but to those of you not involved in transuranium work it might seem strange to be fond of something which one not only rarely handles directly but, in the case of most transuranium elements, also rarely sees. Up to now substantial amounts of the heavier elements (substantial here means more than a few micrograms) have not been readily available. The operation of the facilities we are dedicating today, however, will have much to do with changing all that. And in doing so, it will also bring to Oak Ridge National Laboratory (ORNL) a scientific opportunity unequalled anywhere else in the world.

In a sense the HFIR, the TRU, and the TRL form a triple-threat team for tackling important work in the heavier elements. This team has been over 10 years in the making. Over 10 years ago the Advisory Committee for the Chemistry Division of ORNL pointed out that the Chemistry Division and other parts of this laboratory

Man-made transuranium elements are produced and studied in these special facilities.

could make important contributions to the study of the chemical and nuclear properties of the transuranium elements and that Oak Ridge National Laboratory had many assets for conducting such studies. From 1956 to 1960, the year that the Atomic Energy Commission gave its approval to this reactor, the Advisory Committee, on which I was privileged to serve, encouraged ORNL to acquire a high-flux isotope reactor and the necessary specialized chemical processing capability so that the Chemistry Division and other parts of ORNL could move ahead toward a better understanding of the transuranium elements.

In looking back to the report of the Third Annual Meeting of the Advisory Committee held in October 1956, we find the Committee's first specific reference to the need for the HFIR. The report states: "Every possible encouragement should be given by the division [the Chemistry Division] to the idea of constructing a high flux reactor capable of reaching flux densities of 10^{15} to 10^{16}. This would find many applications and would be especially helpful in producing adequate quantities of transuranium elements for experimental purposes."

In the HFIR control room, (left to right) Alfred L. Boch, Glenn T. Seaborg, and Thomas E. Cole.

Two years later the beginnings of the TRU were forecast in the Advisory Committee's report which said: "Some discussion was held of the proposed construction of a processing laboratory for transuranium elements. The Committee would like to emphasize the importance of this laboratory and to suggest that every effort be made to complete this facility at the earliest date possible."

Many people have worked to make this concept of a transuranium research complex here at Oak Ridge an operating reality. Most of all, credit for organizing and promoting these facilities is due to the late Alfred Chetham-Strode, Jr.

Those of you attending the symposium session this morning heard Drs. Boch, Burch, and Taylor discuss in some detail the attributes of the HFIR, TRU, and TRL facilities. Treating the heavy elements a little more lightly, I will speak only generally about these facilities and what we hope they will accomplish.

The research complex we are dedicating here today will be used to produce and study the transuranium elements—elements avail-

---

Most of all, credit for organizing and promoting the transuranium research facilities is due to the late Alfred Chetham-Strode, Jr. . . .

able on earth only to the extent that they are recreated by man as a result of his scientific investigations and his technological ingenuity. This reactor, this processing plant, this laboratory, and the people who will be working with them will be making a major contribution to our understanding of these new elements, as well as to our overall knowledge of chemistry and physics.

The operation of the High Flux Isotope Reactor is an important step in the AEC's National Transplutonium Production Program, a major aim of which is to make available relatively large quantities of very heavy elements for exploratory research and development. The transplutonium elements produced in this reactor will be used in basic research on nuclear structure and spontaneous fission-decay processes and on heavy-element chemical behavior and biological properties of matter. Although the primary purpose of this reactor is the production of transplutonium elements, it also has several experimental facilities that can be used for research on other problems, such as neutron diffraction and the effects of neutron irradiation on various metals.

(Left to right) A. Chetham-Strode, Catello Franco Cesarano, and G. W. Parker during Cesarano's visit to the hot-cell area of Oak Ridge National Laboratory, Oct. 20, 1958.

A unique fuel assembly
designed here at ORNL. . .

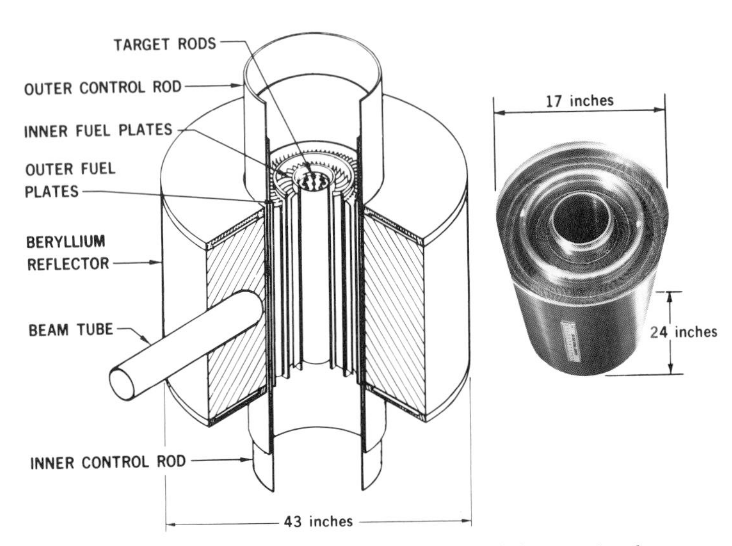

(Top): Central core of the HFIR, designed for use in the production of transplutonium elements. Target rods, shown schematically at the center of the HFIR core drawing, are surrounded by concentric annular regions of fuel plates, control rods, and a beryllium reflector. (Bottom): An irradiated beryllium reflector being examined in a hot cell.

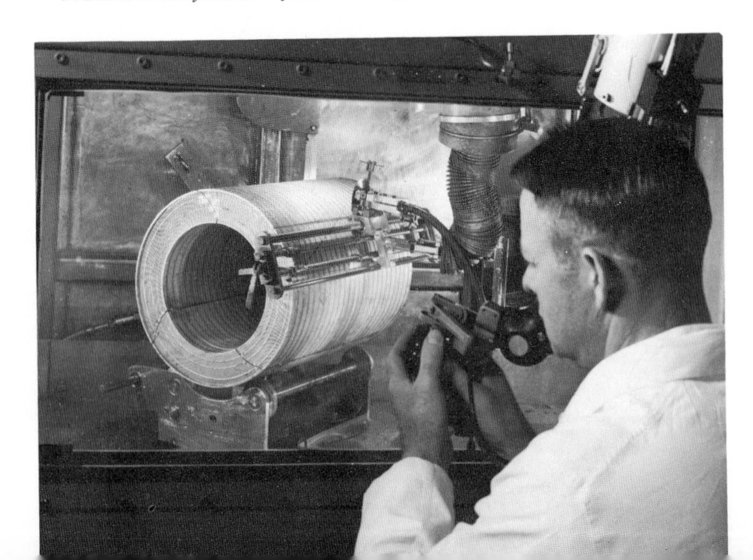

Spent fuel elements from the HFIR in storage under 20 feet of water emit a glow, called the Čerenkov effect, which is caused by high-energy radiation traveling faster than the speed of light in water.

The HFIR is a beryllium-reflected light-water-cooled and -moderated reactor using highly enriched uranium-235 fuel. The design thermal power of the reactor is 100 megawatts, and the maximum unperturbed thermal neutron flux is 5×10^{15} neutrons per square centimeter per second. This high neutron flux is the result of a unique fuel assembly designed here at ORNL. This fuel assembly contains approximately 9.4 kilograms of uranium-235 and 2.2 grams of boron-10.

For use in conjunction with the High Flux Isotope Reactor, Oak Ridge National Laboratory will have the Transuranium Processing Plant, another unique transuranium facility. This plant will chemically isolate and separate the higher transuranium elements from the target material that is irradiated in the High Flux Isotope Reactor. It also will prepare the separated transuranium elements to be reirradiated by the reactor since recycling over a long period of time is required to produce the expected quantities of the heaviest transuranium isotopes.

The heavy isotopes produced by this reactor and this processing plant will be available for research here at Oak Ridge and at other laboratories engaged in transuranium research throughout the United States and other countries. The annual recovery of isotopes from HFIR is expected to amount to grams of californium, hundreds of

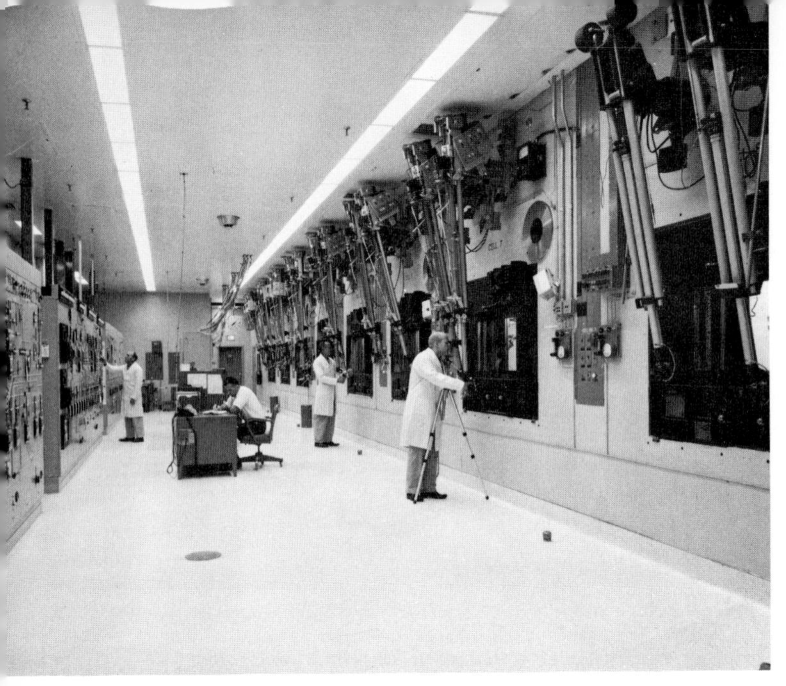

In the hot-cell area of the Transuranium Processing Plant, target rods are prepared for irradiation and for separation of elements produced by irradiation.

Production of heavy transuranium elements from plutonium-239. This chart illustrates the major path of transuranium-element production by successive neutron captures in a nuclear reactor. Both fission and neutron capture take place in plutonium-239 nuclei. About 70% fissions and is lost; the remainder is transmuted to plutonium-240. The plutonium-240 nuclei then capture neutrons and become plutonium-241. The split between fission and capture is repeated at all the even–odd isotopes up the chain so that only 0.3% remains as transuranium elements when californium-252 is reached.

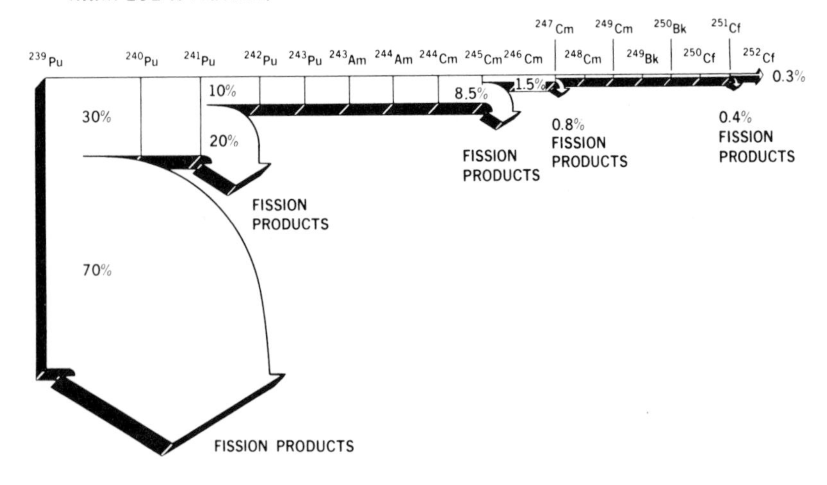

Recycling in the HFIR over a long period of time is required to produce the expected quantities of the heaviest transuranium isotopes. . .

---

milligrams of berkelium, tens of milligrams of einsteinium, and micrograms of fermium.

The reactor and the processing plant will serve another of the outstanding scientific resources we are dedicating here today: the Transuranium Research Laboratory. This unique laboratory has special facilities and equipment for making chemical and nuclear studies of the highly radioactive transuranium elements. Because of their rapid decay, many of the transuranium elements will be available only at Oak Ridge, and scientists from other laboratories in

The plutonium is irradiated at the Savannah River Plant. The plutonium-242 thus formed is isolated and shipped to Oak Ridge along with a mixture of americium–curium and rare-earth fission products. The plutonium is made into targets for the HFIR without further treatment. The americium–curium is separated from the rare earths in the TRU and made into HFIR targets. After exposure in the HFIR for about 18 months, the targets are returned to TRU, where the transuranium elements are isolated and supplied to researchers. The curium isotopes are returned to the HFIR for further irradiation to produce still more heavy transuranium elements.

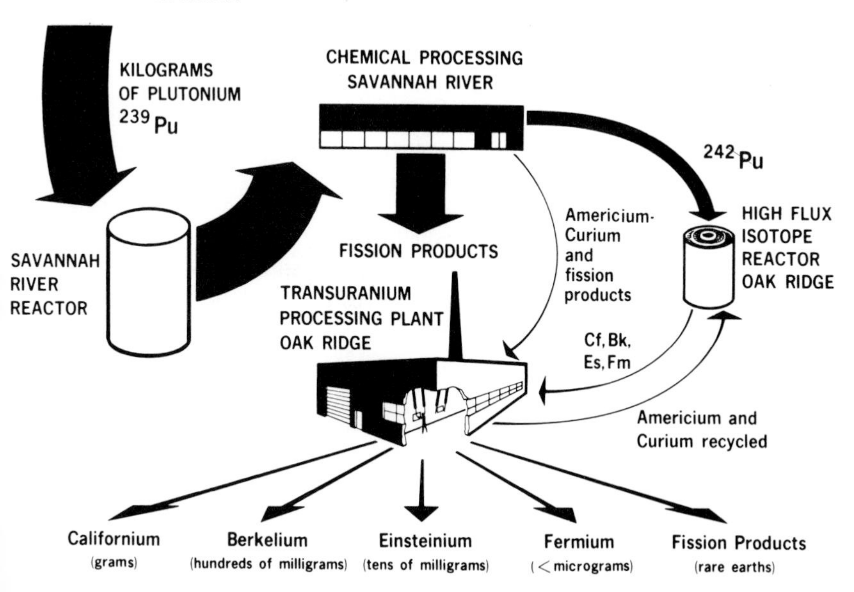

the United States and foreign countries will join ORNL staff members in performing research on the heavy elements at the Transuranium Research Laboratory.

The knowledge gained from experiments on the transuranium elements will deepen our comprehension of nature by increasing our understanding of atomic and nuclear structure. This is the primary reason for the intricate and exotic experiments that will be made possible by the facilities we are dedicating this afternoon. It is, of course, not possible to indicate the course of such basic research in any detail.

It will be possible to prepare isotopes with relatively high isotopic purity, such as californium-249, with the help of chemical isolation and decay of radioactive precursors. The judicious employment of neutron irradiation of intermediate products will also allow the preparation of numerous heavy isotopes with relatively high isotopic purity. These will be very useful in studying the chemical properties of the heaviest transuranium elements, in studying the nuclear properties of numerous heavy isotopes to a degree hitherto not possible, and as target material for the preparation of additional isotopes through bombardment with neutrons or charged particles. The time may come when the facilities here should be augmented by additional facilities, such as electromagnetic separation apparatus, for the preparation of individual transuranium isotopes in isotopically pure form.

As has often developed from the results of basic research, there will be practical applications with far-reaching impact. Some of these applications have already appeared, and many others are sure to follow.

The enormous practical importance of plutonium in the form of the fissionable isotope plutonium-239 is, of course, well known. This material can be used not only as the explosive ingredient for nuclear weapons but also as a nuclear fuel to generate electricity to serve the world's needs for centuries to come.

Not so well known are the potential practical uses of other transuranium isotopes. These presumably would require production facilities with capabilities beyond these research facilities at Oak Ridge National Laboratory.

An example is the isotope plutonium-238, which requires very special production capabilities. This isotope, which has a 90-year half-life and was the first isotope of plutonium to be discovered, may prove to be one of the most valuable assets to mankind. It can be used as a compact source of electricity through the conversion of its heat from radioactive decay by thermoelectric or thermionic devices. Such plutonium-238-fueled power units are very compact and light in weight and hence are admirably suited for long-lived power sources in space and terrestrial applications. It is important to note that such plutonium-238 power sources have one very special advantage: they can be handled safely and directly by technicians since they emit very little external radiation.

One potential terrestrial use of plutonium-238 is to power pacemakers for heart patients. Even more exciting is the idea now being considered of using plutonium-238 to power an entirely artificial heart that can be surgically implanted in the patient. The projected requirements for plutonium-238 in space nuclear batteries over the next decade or two run into tons of material. If the artificial heart application should materialize, the requirements will be substantially greater than this.

The possible practical applications of other transuranium isotopes are even more interesting to me. For example, it is possible to produce curium-244, an isotope with an 18-year half-life, by the intense neutron irradiation of plutonium. This isotope, like plutonium-238, also can be used as a fuel for nuclear batteries.

I have a special interest in the large-scale production of curium-244 because of the tremendous by-products of the heavier transuranium isotopes that would inevitably result from its production in large amounts. This would constitute a production capability for such heavier transuranium isotopes as a follow-on to the HFIR. If curium-244 turns out to have the properties required for widespread use as an isotopic power source, one can imagine its future production in ton quantities. This could be accomplished, for example, by the conversion of a number of plutonium production reactors to the manufacture of curium-244. Such quantities of curium-244 are almost beyond comprehension when we recall that our first experience with this isotope, as recently as 1950, involved picogram quantities (that is, quantities of the order of a million-millionth of a gram). Now we are preparing a 3-kilogram batch of this isotope at our Savannah River Plant in order to make an assessment of its properties as an isotopic power source! Thus we are now talking about an escalation in production of a million billion fold.

As an alternative to producing larger quantities of heavier transuranium elements as by-products of the large-scale production of curium-244, one could, of course, produce such isotopes as primary objectives in large reactors to meet the possible demand for their potential applications. One such isotope is californium-252, which is produced through successive neutron capture in high-flux nuclear reactors. This isotope will be produced in the HFIR in quantities sufficient to assess its practical applications. Because of its high rate of neutron emission, it has potential widespread applications as a point neutron source for radiography or as a portable and reliable source for conducting neutron-activation analyses in locations on earth and out in space where it is not possible to use conventional neutron generators. It may also have widespread applications for medical therapy.

The applications already discovered or presently foreseen for the heavier transuranium elements are among the most promising frontiers of contemporary science and technology. But I am sure that there will be many more important applications that are beyond our immediate vision—applications that will not only benefit mankind's health and welfare but also deepen our basic understanding of the natural universe in which we live.

Today chemistry, physics, and biology are probing the very basis of our universe, giving us a better understanding of the beauty, order, and logic of nature. But discoveries in these fields are also providing us with a new mastery of nature and imposing upon us the responsibilities that go with this mastery. Research into the transuranium elements led us to discover 10% of those elements now in the periodic table. During the past 25 or 30 years this research has stimulated a veritable renaissance in the field of inorganic chemistry. Among the lessons to be learned from our adventures in today's science is that knowledge—all knowledge—is worth pursuing to the utmost. The quest for basic knowledge must continue for its own sake, because the desire to know is one of the things that make us human. In addition, we realize that all knowledge, sooner or later, can have a significant impact upon our lives.

You who are privileged to work in this great national laboratory are working close to mankind's newest frontiers. It is partly upon the legacy of your pioneering work that others will build the world of the future. As scientists in this age of science, we all have a great responsibility to our fellowman. In this respect the scientific research facilities we are dedicating today are a testimony to our country's faith in the future and to the importance of your role and influence in the years ahead. As modern science has demonstrated, the desire for understanding increases with the acquisition of it.

So I can say with confidence that in the years ahead one of the most stimulating and exciting subjects of scientific research will be the transuranium elements. I am equally confident that the facilities and the people at Oak Ridge National Laboratory will have a very distinguished future in transuranium research. I congratulate you on the dedication of these new facilities. ■

NATIONAL ACCELERATOR LABORATORY

At the ceremony, Robert R. Wilson, director of the National Accelerator Laboratory, and Glenn T. Seaborg.

Ground-Breaking Ceremony for the National Accelerator Laboratory, Batavia, Illinois, Dec. 1, 1968

■ This is a very exciting and meaningful day for all of us gathered here. With this ground-breaking ceremony we move one step closer to the realization of a great new scientific enterprise—a national laboratory that will help man to advance significantly into new frontiers of knowledge. Symbolically, we could say that the spade that breaks ground on this site today begins our deepest penetration yet into the mysteries of the physical forces that comprise our universe.

Since this is an historic event, perhaps it is appropriate that we recall some of the historical developments culminating in this ground-breaking ceremony here today. When we reflect on the relatively short history of high-energy physics, there can be no doubt that the development of this important area of fundamental scientific research has been astounding in terms of productivity and growth. Only a little more than 20 years ago plans were initiated for constructing accelerators in the billion volt range. These early plans materialized in the Cosmotron at Brookhaven and the Bevatron at Berkeley; both truly prodigious in their contributions to the field. As time went on other excellent machines and facilities were designed and built, and their exploitation has assured the solid growth and progress of high-energy physics. This strong dependence on large, complex, and costly machines inevitably brought wide national attention to the field.

Throughout the history of the 200-BeV accelerator project the interest and encouragement given by the Joint Committee on Atomic Energy has been crucial. The committee's concern for scientific progress, along with essential program economies, provided a firm foundation from which to move ahead with the project. In its report for fiscal year 1965, the committee strongly urged that the Atomic Energy Commission take the lead in developing a national policy for the high-energy-physics programs. The Commission responded by developing the document entitled "Policy for National Action in the Field of High-Energy Physics," which was transmitted to the President and then by him to the Joint Committee on Atomic Energy in January 1965. The policy for this field of research had been evolving over a period of years, marked by a series of studies and reports beginning in 1954.

All these studies contributed to the formulation of the document prepared by the AEC; indeed the text of each of these reports is embodied in that policy document. After the Joint Committee received the policy statement from the President, its Subcommittee on Research, Development, and Radiation, under the chairmanship of Melvin Price, held a week-long series of hearings in March 1965 on the status and achievements of high-energy physics, including discussions of the 200-BeV accelerator and the policy document.

Let me read to you a portion of the document:

> Proton energy is the single most important parameter to be extended. . . .
>
> There are now clear needs for a proton machine following the conventional alternating gradient synchrotron (AGS) design but having an energy of hundreds of BeV. These needs can best be met by the immediate design and construction of an AGS in the 200 BeV range. . . .

The Lawrence Radiation Laboratory of the University of California spent two years of intensive design work on an accelerator in this range, and in April of 1965 the Commission began one of its longest, most interesting, and certainly most difficult tasks—the selection of the most appropriate site in the United States for a 200-BeV proton accelerator. Site criteria for such a national laboratory for basic research were issued; they were based on studies done by the Lawrence Radiation Laboratory. The avalanche of site proposals began, and for the next 20 months the Commission, its staff, and a select committee appointed by the National Academy of Sciences and headed by Emmanuel Piore wrestled with the detailed evaluation of 126 proposals that included well over 200 sites in 46 states.

On-site visits by inspection teams were made to the nearly 150 sites that met the basic criteria, and by March 1966 the National Academy of Sciences Site Selection Committee recommended to the Commission six areas—actually seven different sites. Two sites were in the Chicago area.

On Dec. 16, 1966, the Commission, after many meetings and intensive review of all the material on the recommended sites, unanimously selected this site for the 200-BeV accelerator. It was a busy day, and there have been some busy days since then.

Meanwhile, another very important aspect of the new project was under consideration, namely, the management of the project. From the beginning it was understood that the 200-BeV accelerator would be a national facility available to all on the basis of scientific merit. It was also clear from the outset that the management of the facility must have national orientation and representation.

It was not at all clear, however, how this was to be accomplished.

Many discussions were held on this matter. Early in 1965 Frederick Seitz, after consultations with the Office of Science and Technology, the AEC, and the National Science Foundation, called together representatives from universities most directly concerned with high-energy physics to consider appropriate managerial patterns for future high-energy nuclear accelerators. At this meeting held at the National Academy of Sciences on Jan. 17, 1965, it was agreed that the universities should form an association, nationwide in membership, for the purpose of offering its services as manager of future large federal research facilities. The resulting organization, now known as Universities Research Association (or URA), was organized initially with 34 members under its first president, J. C. Warner. He was succeeded by the current president, Norman F. Ramsey, and the organization has grown to 48 members including one in Canada.

I would like to turn for a few moments to another most important aspect of this project. Early in 1967, soon after the Commission had announced that the National Accelerator Laboratory was to be built here, many questions were asked concerning the civil-rights aspects of the selection. The Commission was severely criticized by some for having selected a site in an area which did not appear to have accepted the concept of equal opportunity for all regardless of race, creed, color, or national origin. In answer to this criticism, I told the Joint Committee on Feb. 8, 1967, that "a satisfactory solution to the human rights problem is more important than this accelerator. However, they are not in conflict here at the Weston site. We believe that construction of the accelerator at Weston and advancement of human rights can complement one another." In short, we regard the civil-rights challenge as an opportunity to be met, and I believe that we are meeting that opportunity in the right way.

The National Accelerator Laboratory has been a catalyst for change. No less than 14 communities within the commuting radius of this site have adopted open occupancy laws since the selection was announced.

Indeed, NAL has been far more than a catalyst. It has actively and effectively promoted policies of equal opportunity looking both to the present and the future, in cooperation with AEC's Chicago Operations Office and Daniel, Urbahn, Seelye, and Fuller, the architect–engineer. It has cooperated with local unions in promoting preapprenticeship and other training programs for the construction industry and in recruiting minorities to participate in these opportunities. Thus far 71 young minority men have been given a chance, previously unavailable to them, to work in the construction industry in this area. In promoting these programs NAL and its associates have found cooperation and even encouragement from local building trades unions, from local industry, and from local community leaders.

Both in its construction and in its operation, the laboratory will continue to provide such opportunities for disadvantaged men and women in this area. The dedication to equal opportunity shown by the laboratory and its associates will, I believe, demonstrate that with enlightened federal, state, and industry cooperation scientific progress and human progress can go hand in hand.

At the beginning of 1967 the Atomic Energy Commission entered into its first contract with the Universities Research Association for

Aerial view of the "injection area," April 1971. Protons are generated in the Cockcroft–Walton Accelerator, then introduced into the Linear Accelerator (Linac). At 200 MeV, the protons are injected from the Linac into the Booster Accelerator which carries them to an energy of 8 BeV. At 8 BeV, the protons are injected into the Main Accelerator ring (part of the covered ring is visible to the right) and are here accelerated to full energy.

Inside the 500-foot-long Linear Accelerator tunnel. Several of the nine proton-accelerating tanks can be seen.

Inside the Booster Accelerator tunnel, a ring 500 feet in diameter containing 96 magnets.

The mile-and-a-quarter diameter Main Accelerator ring. . .

Aerial view of the Main Accelerator ring at the National Accelerator Laboratory. In foreground at right is the Central Laboratory area including the Preaccelerator, Linac, Booster, Beam Transfer, and Central Laboratory building (under construction).

Bending magnets inside the Main Accelerator ring.

Precast concrete tunnel sections of the Main Accelerator ring enclosure being placed. The ring has a circumference of about 4 miles and required about 2000 sections, each 9 feet high and 10 feet long. The entire tunnel is now covered by earth shielding.

It was 5:28 p.m. Friday afternoon, Apr. 16, 1971, when the 1,014th magnet — the final one—was put in place to complete the 4-mile Main Accelerator ring. Here, Robert R. Wilson, NAL's director, is being congratulated by Chairman Glenn T. Seaborg, U. S. Atomic Energy Commission, on the accomplishment.

Congratulations also from visiting Andronik M. Petrosyants, Chairman, USSR State Committee on Atomic Energy.

a design study, and the National Accelerator Laboratory came into being. The first step was to find a director; happily the choice was Robert R. Wilson, who was then completing work on a large electron accelerator at Cornell. He rapidly assembled a design group at Oak Brook, in the western suburbs of Chicago. In 1967 a summer design effort involving many of the nation's accelerator specialists took place. An NAL Users Group reaching into almost every institution in the country where theoretical or experimental high-energy physics is studied was organized the same year to assist the laboratory in planning.

Meanwhile, the staff of the NAL was taking form, and in the current year their varied efforts and substantial achievements have provided a very impressive beginning. The signing of a definitive design and construction contract last January and the appropriation only a few months ago of funds to begin construction of the laboratory's first buildings assure that progress is to continue.

During all this time, from the beginning and through all these important stages of planning and initial work, we have had the firm support of President Johnson for this project. And we are grateful for that support.

Now we have come to the point of breaking ground for the first of the buildings to house the accelerator. Appropriately enough, the first structure to be built is the housing for the linear accelerator. Eventually, in this housing the protons will begin their trip through the accelerator.

This linear-accelerator housing is the beginning of the major construction program aimed at completing the accelerator itself by mid-1972 and the laboratory as a whole by the end of 1973. After the construction of this housing is started, the booster-accelerator housing will be initiated, followed by the mile-and-a-quarter diameter main ring tunnel. At the same time, construction of the accelerator components—magnets, vacuum chambers, power supplies, and so forth—will have been started, and, as the accelerator housings are completed, the accelerator systems will begin to take shape within them.

The linear accelerator will be completed first. Then, when the booster accelerator is finished, its shakedown will be carried out using protons injected from the linear accelerator. Likewise, when the main ring accelerator is ready to receive protons, they will be available from the booster. Thus each component of the 200-BeV accelerator complex will be tested on a timely schedule.

This construction schedule is an intensive and demanding one. The payoff will come, if all goes well, in the middle of 1972 when

Professor Wilson and his staff expect to achieve the first 200-BeV proton beam and to begin the nation's, and the world's, first controlled multihundred BeV research program.

Particle accelerators have long been indispensable to nuclear research as the sources of energetic particles with which to probe the nucleus and its constituents. The Atomic Energy Commission has sponsored a series of high-energy accelerators for the investigation of the fundamental particles and laws of physics. These accelerators, the Cosmotron and Alternating Gradient Synchrotron at Brookhaven National Laboratory, the Bevatron at Berkeley, the Zero Gradient Synchrotron at Argonne National Laboratory, the Stanford Linear Accelerator, the Cambridge Electron Accelerator, and the Princeton–Pennsylvania Accelerator, have produced a wealth of new knowledge: the verification of the existence of antiparticles of nucleons, the discovery of a host of new particles, the discovery of unexpected asymmetries in nature, and many other things. Each time a new regime of energy has been entered a wealth of new phenomena has been revealed, and we have come closer to an understanding of the forces at work in the nucleus.

With the 200-BeV accelerator at the National Accelerator Laboratory, we take another step in energy, eagerly anticipating the revelations the new energy regime may have in store for us.

We are proud of the accomplishments of the URA, of the National Accelerator Laboratory, and of all the distinguished scientists and engineers who have contributed to this project. They have produced an efficient and economical accelerator design that has favorably impressed scientists all over the world. They have planned and initiated an expeditious construction schedule. This ground-breaking ceremony this afternoon symbolizes the start of the major construction of the National Accelerator Laboratory. We congratulate Professor Wilson and his staff, and we look forward to a highly successful program. ■

Index

Photo Credits

PAGE	SOURCE
2, 4, 9, 10, 11 (middle and bottom), 14, 21, 25, 27, 40–42, 52, 53, 70, 82, 84, 85, 88 (top and bottom), 89, 90 (top and bottom), 91, 93, 94, and 99	Lawrence Radiation Laboratory
5 and 74 (bottom)	Herbert Weitman, Washington University
11 (top)	Philip H. Abelson
15, 29, 31 (top and bottom), 33, 34, 35 (top and bottom), 36, 37, 39, 57, 77 (bottom), 80, 83, 97, 128, and 129	Argonne National Laboratory
23, 30, 74 (top), 79	University of Chicago
38	Chicago Daily News
45, 136 (bottom)	Oak Ridge National Laboratory
50	Martin Marietta Corp.
72	World Wide Photos
77 (top)	U.S. Army
109	Alan W. Richards
100 (top, middle, and bottom), 101 (top), 102, 105, 110, 132, 134, and 135	Los Alamos Scientific Laboratory
122	Addison-Wesley Publishing Company, Inc.
141	U.S. Atomic Energy Commission
154, 178, 360, 361, 362, 363, and 364 (top)	Oak Ridge National Laboratory

155, 156, 157, 158, 159 (top), 164, 168, 169, 170, 171 (top), and 172	U.S. Army
162	Pacific Northwest Laboratory
165	Robley L. Johnson
171 (bottom), 204, 205, 210, 249, 307, and 338 (bottom)	Ernest Orlando Lawrence Berkeley Laboratory
174 and 175	General Electric Company
180, 182, 183 (top), 184, 185, 187, 188, 189, 190, 191, 192, 193, 194, 196, 197, and 198	Los Alamos Scientific Laboratory
186	Donald W. Kerst
200, 201, 202, 208, and 213	Ernest Orlando Lawrence Livermore Laboratory
203	Harlingue, The Niels Bohr Library
216, 219, 220, 304, 305, 308, 312, 315 (right), 318, 319, 320, 321, 322, 326, 327, 328 (bottom), 329, 331, 333, 334, 335, 336, 337, 338, 341, and 370	Argonne National Laboratory
224, 225, 226, 228, 230, 231, 234, 235, 236, 245, 246, 247, 250, 251, 252, 254, 255, 256, 257, 260, 261, 262, and 263	Brookhaven National Laboratory
264, 268, 269, 270 (top), and 271 (bottom)	Ames Laboratory
274, 275, 277, 280, 282 (top), 284, and 285	Monsanto Research Corporation, Mound Laboratory
286, 288, 289, and 290	Stanford University
305 (left)	Spofford G. English
314	University of Chicago
336, 338 (top), 341 (bottom), 344 (top), 345 (top), 349, 350, and 351	Savannah River Laboratory
341 (top) and 345 (bottom)	Morgan Fitz
374, 375, 376, 377, and 378	National Accelerator Laboratory